Ultran

LABORATORY MONOGRAPHS

Malignant Blood Diseases
Tissue and Cell Culture
Servicing Electronic Laboratory Equipment
Ultracentrifugation
Laboratory Planning

Ultramicrotomy

R. L. Griffin
F.I.M.L.T.

General Hospital
Southampton

BAILLIÈRE TINDALL

7 & 8 Henrietta Street, London WC2 8QE

A division of Crowell Collier Macmillan Publishers Ltd.

ISBN 0 7020 0426-X

Published in the United States of America by
The Williams & Wilkins Company, Baltimore

Printed in Great Britain by
The Whitefriars Press Ltd., London and Tonbridge

Preface

Meticulous examination of ultrastructural detail is almost certainly the most used method of studying tissue. It can be done only when extremely thin sections have been cut from specimens which have been fixed and then embedded in media suitable for electron microscopy; ultramicrotomy is probably the most difficult single technical procedure involved.

The ultramicrotome, an instrument designed specifically to cut thin sections has greatly increased the rate at which good specimens can be prepared. Knives can be made by hand or with a knifemaking machine and both methods give a high rate of success in the hands of a skilled operator.

In this book I have described in detail methods which I have found to be reliable. I recommend the beginner to use them until his technique is sure; he can then consult the literature for other methods appropriate to his laboratory's needs.

The novice must approach ultramicrotomy in the knowledge that it demands a great deal of patience and attention to detail. A cavalier approach may occasionally yield good sections but only a methodical, standardized technique can produce high quality work as a matter of routine.

February 1972 R. L. Griffin

Contents

Introduction

Before the ultrastructural detail of biological tissues can be demonstrated in the electron microscope, the tissue must be suitably prepared. The main stages in the preparation of tissues are analogous to those conventionally used in the preparation of tissue for the light microscope, but the reagents and materials are different. The main steps consist of fixation, embedding, ultrathin sectioning and staining.

FIXATION – The purpose of fixation is to preserve tissues and cellular detail in as near a life-like condition as possible and to arrest natural decomposition. At the high magnifications used in electron microscopy, any failure of the fixative to fix the tissue as near as possible to the ideal is easily seen and neccessitates the use of specialized fixation techniques.
Note: Osmium tetroxide (OsO_4 molecular weight 254.2) is the tissue fixative most commonly used in electron microscopy: it is also known as osmic acid. I have used this term when writing of osmium tetroxide in aqueous solution.

EMBEDDING – In order to cut the very thin sections of tissue required by the electron microscope, the tissue is normally impregnated with and embedded in a hard, sectionable material. This material, usually a synthetic resin, is then made to solidify. In order to infiltrate the tissue with the resin, all water is removed from the tissue block by a process of dehydration, normally achieved with increasingly stronger solutions of alcohol.

ULTRATHIN SECTIONING – To visualize the maximum detail of sections in the electron microscope, sections of the order of 30-80nm thick are required. Ultramicrotomes, specialized instruments using glass or diamond knives, have been developed to cut these very thin sections.

STAINING – The ultrathin sections are stained with heavy-metal salt solutions to increase the contrast of structures within the section, making them clearly visible.

Acknowledgements

I wish to acknowledge the constructive criticism of the text given by Dr. M. A. Woodhouse and to express my appreciation to Marjorie Blackman and Maureen Paterson for typing the manuscript.

I also wish to thank my wife Daphne for her assistance with the typing and her helpful encouragement during the preparation of the book.

1 Tissue Fixation

In common with other branches of histology, the results obtained in electron microscopy depend upon the quality of the fixation of the tissue. Whilst a reasonable standard can be achieved with post mortem tissue and large surgical specimens fixed in formol saline, perfect results can only be obtained from small human or animal biopsies using specialized fixation techniques.

The fixatives most commonly used contain either buffered osmium tetroxide or an aldehyde, for example glutaraldehyde or formaldehyde. Potassium and sodium permanganates have also been used as fixatives for electron microscopy; but they have failed to gain favour in routine use for animal tissue, possibly because of difficulties in sectioning blocks of tissue fixed in them. These fixatives are buffered with a variety of agents. Veronal acetate, phosphate, cacodylate or S-collidine buffers are usually employed.

The cytological detail will vary slightly in a given tissue depending on the particular fixative used. The choice of fixative may have to be a compromise between the dictates of the problem under investigation and the technical difficulties.

OSMIC ACID FIXATION

Osmium tetroxide is a pale yellow, crystalline solid with a characteristic odour. The toxic effects of osmic acid vapour, even arising from solutions, can cause inflammation of the nasal mucosa and sometimes a severe conjunctivitis. These effects can be avoided by working in a fume cupboard. If this is inconvenient a strong stream of air from an electric fan or open window may be used to disperse the vapour so that the user does not come into contact with a high concentration.

As a fixative, osmic acid is generally used as a 2 per cent buffered solution. It acts very rapidly on the cells near the surface of tissues, but its penetration is poor. Even with prolonged fixation, it only penetrates to a depth of approximately 0.25 mm. To ensure that the centre of the block is well fixed, its overall dimensions should not exceed 1 mm. Used under the right conditions, osmic acid fixatives give excellent cytological detail and increase the contrast of ultrathin sections, an added advantage for electron microscopy.

GLUTARALDEHYDE FIXATION

Glutaraldehyde is a clear, colourless liquid, supplied commercially as a 25 per cent solution in water. For use in electron microscopy, this solution is diluted with a phosphate or cacodylate buffer of pH 7.2-7.4 to give a final glutaraldehyde concentration of 4-6 per cent. Difficulties may be encountered with this fixative because the pH of the stock 25 per cent solution is liable to fall fairly quickly to lower than 3.5. When the pH of the stock solution falls to this level and below, poor fixation follows. Glutaraldehyde may have a lower pH than this when first received from the supplier and may have to be rejected. This problem can be avoided by purchasing glutaraldehyde purified specially for electron microscopy and by discarding the stock when the pH reaches 3.5.

Glutaraldehyde gives good preservation of the fine cellular structure and preserves more cell protein than osmic acid. This is noticeable in the increased amount of chromatin visible in cell nuclei and the increased density of the cytoplasm of cells. It has the added advantage that electron microscope histochemistry can be performed, even at a later date if desired. The size of the block to be fixed in glutaraldehyde is not as critical as with osmic acid. Pieces of tissue up to 3 mm thick can be fixed without loss of quality. This is of considerable convenience in pathology laboratories where a single piece of tissue may have to be shared between several workers after fixation.

FORMALDEHYDE FIXATION

A 40 per cent solution of formaldehyde (free from methanol) in water is diluted with a cacodylate or phosphate buffer at pH 7.2-7.4 to give a fixative in which the final concentration of formaldehyde is 4 per cent. Because formaldehyde destroys the buffering action of veronal acetate, the two cannot be used together. Formaldehyde fixatives tend to produce more shrinkage and generally give less satisfactory preservation of cell ultrastructure than glutaraldehyde.

POST FIXATION IN OSMIC ACID

Due to the failure of aldehyde fixatives adequately to protect some lipids from the solvent action of embedding fluids, tissues fixed in aldehydes require an additional fixation in osmic acid of one to two hours in order to make them visible in the electron microscope. This post fixation, often referred to as post-osmication, is normally carried out when all of the original aldehyde fixative has been removed by soaking the tissue in a suitable buffer.

COMPOSITION OF FIXATIVES

Osmic acid (Palade, 1952)

This fixative is made by adding osmium tetroxide crystals to Michaelis veronal acetate buffer. It can be used with confidence as a fixative for a wide range of tissues for electron microscopy.

Preparation of Michaelis veronal acetate buffer stock solution

Barbitone sodium	2.94 g
Sodium acetate (anhyd.)	1.15 g
Distilled water	100 ml

This buffer keeps for several months at 4° C.

Preparation of 2 per cent osmic acid solution (50 ml volume)

Glass ampoules containing 1.0 g of crystalline osmium tetroxide are normally supplied with a gummed label on the outside of the ampoule. This label and all traces of glue must be removed by soaking the ampoule in water until the label floats off. Wash the ampoule well in distilled water, then transfer it to a clean glass-stoppered bottle with clean forceps. The bottle should not be rinsed in alcohol as part of its cleaning process. These precautions are necessary due to the reduction of osmic acid when it comes into contact with organic contaminants such as glue, paper, alcohol, finger prints etc.

In order to break the ampoule, shake the bottle vigorously. Alternatively, a strong, clean glass rod approximately 10 mm in diameter can be introduced into the neck of the bottle and used to smash the ampoule. This method may appear to be a little crude, but in practice it is safe, effective and convenient.

When the ampoule has been broken, add the following buffer solution, which should have been prepared previously:

Veronal acetate buffer stock	5 ml
0.1 M-Hydrochloric acid	5 ml
Distilled water	20 ml

Adjust the pH of this solution to the range of 7.2-7.4. Add a drop at a time of either the veronal acetate stock solution or the 0.1 M-hydrochloric acid, depending on whether the pH is above or below that required. In practice, only a few drops of either buffer or acid will be required. Monitor the pH with a pH-meter. If this is not available, use narrow-range indicator papers. Make up to 50 ml with distilled water. The crystals normally take two or three days to dissolve at 4° C.

It is possible to purchase ampoules of dispersed solid osmium tetroxide in which a finely dispersed crystalline layer is spread over the whole of the inside of the glass surface. In this form, the time necessary to make the fixative is reduced because it dissolves more rapidly.

Provided that precautions are taken to avoid contamination of the fixative with foreign matter, the buffered solution of osmic acid, ready for use, will keep for several months at 4°C. The solution is normally a pale yellow colour. If it turns black, it should be discarded because it has been reduced and is no longer a satisfactory fixative.

Glutaraldehyde fixative

This is made by buffering stock glutaraldehyde solution with a 0.2 M aqueous solution of sodium cacodylate $(CH_3)_2AsO_2Na.3H_2O$.

0.2 M aqueous sodium cacodylate	33 ml
Stock glutaraldehyde	20 ml
Distilled water	47 ml

Adjust the pH to 7.2-7.4 with 0.1 M sodium hydroxide or 0.1 M hydrochloric acid. The fixative is stable for several weeks at 4°C. If only small quantities are being used the remainder should be discarded and fresh fixative made up at monthly intervals.

Fix small pieces of tissue up to 3 mm thick for 4-6 h at 4°C; larger pieces need a proportionally longer time. When fixation is complete, tissues are transferred to the sucrose/cacodylate buffer, described below, in which they can be stored indefinitely at 4°C. The tissue should be washed in the sucrose/cacodylate buffer for a period lasting at least twice as long as the original fixation if, after fixation in glutaraldehyde, the specimen is to be post fixed in osmic acid or cytochemistry is to be undertaken. Normally, a maximum time of 24 h is sufficient.

Sucrose/cacodylate buffer

0.2 M Aqueous sodium cacodylate	100 ml
Sucrose	7 g

Adjust the pH to 7.4 with 0.1 M sodium hydroxide.

Formaldehyde fixative

This is made by adding formaldehyde solution to cacodylate buffer.

0.2 M Aqueous sodium cacodylate	33 ml
40 per cent formaldehyde solution (methanol free)	10 ml
Distilled water	57 ml

Adjust the pH to 7.2-7.4 with 0.1 M sodium hydroxide or 0.1 M hydrochloric acid.

Fix small pieces of tissue up to 3 mm thick for 1-2 h at 4° C, and give the same post fixation treatment as glutaraldehyde.

Potassium permanganate fixative (Luft, 1956)

This is made by adding an aqueous solution of potassium permanganate to Michaelis veronal acetate buffer.

Solution A

Michaelis veronal acetate buffer pH 7.2-7.4 as used for osmic acid.

Solution B

A freshly prepared 1.2 per cent solution of potassium permanganate in distilled water.

For use, mix equal parts of A and B. Fix in this mixture at 4° C. The time necessary is best found by trial and error but is approximately 1-2 h, depending on the tissue.

TAKING TISSUE FOR ELECTRON MICROSCOPY

Tissues for electron microscopy must be taken as soon as possible after the blood supply has been interrupted, and they must be placed into fixative with an absolute minimum of delay. The method is as follows:

1. Slice off a piece of tissue approximately 1 mm thick from the main part of the specimen as soon as possible after removal

from the body, and place it immediately into a puddle of osmic acid fixative chilled to 4° C on a block of paraffin wax. Use a new grease-free razor blade to cut the sample; scalpels are not sharp enough. Remember the hazards of osmium tetroxide vapour.

2. With a cutting action, dice the tissue into 1 mm cubes with a razor blade. The ultrastructure of the tissue will be damaged by using a pressing rather than a cutting action.
3. Transfer the blocks of tissue to a small specimen bottle containing 1-2 ml of fresh fixative at 4° C.
4. Fix the tissue for 2 h at 4° C.
5. Discard the fixative using a Pasteur pipette, and replace it with distilled water. Leave for 2 min. Swirl the blocks of tissue around in the distilled water.
6. Change to 70 per cent ethanol. The tissue can now be stored in 70 per cent alcohol or embedded.

This method is very useful for tissues obtained from experimental animals or human biopsies.

In the case of tissue being taken from experimental animals, superior fixation may be produced by dripping osmic acid fixative onto the cut surface of the tissue *in situ* in anaesthetized animals for 15-30 min before the tissue is removed. This technique produces excellent results in the superficial cell layers, but the poor penetrating power of osmic acid must be remembered. Again a compromise between dripping and total immersion of a tissue partly fixed *in situ* is often necessary, depending on the tissue under investigation.

In the case of the central nervous system, perfusion of the whole body of the anaesthetized animal with buffered glutaraldehyde has been found to give the best preservation. The reader should refer to the paper by Gonzalez and De Robertis for details of this technique.

CRITERIA OF FIXATION

The following points are given as a guide to good fixation with osmium or glutaraldehyde. The nuclei of tissue fixed in osmic acid should exhibit a fine, uniform granularity, unlike nuclei fixed by glutaraldehyde which show chromatin masses, sometimes in a characteristic manner. For example, plasma cells exhibit clumping of chromatin at the periphery of the nuclei in a particularly dramatic way. Mitochondria are very sensitive to fixation; if they are seen to be abnormally swollen or ballooned, this may indicate poor fixation. Other indications of poor fixation may include vacuolated cytoplasm, distended cisternae of the endoplasmic reticulum and disrupted cell membranes. Pathological changes can mimic artefacts produced by poor fixation, and this must be borne in mind.

For purposes of comparison, examples of very well fixed tissues may be found in the numerous excellent electron micrographs in a number of published atlases of ultrastructure.

Although good fixation is always desirable, the examination of tissue removed at necropsy may be profitable, provided that the actual details under investigation have not been altered by autolysis. Great care must be exercised in the interpretation under these circumstances. However, the localization of small amounts of extra- and intracellular electron-dense substances (for example silver, iron, calcium salts and virus particles) can be demonstrated with a reasonable degree of accuracy.

2 Embedding

Embedding media available for electron microscopy include methacrylates, epoxy resins and polyester resins. Methacrylates were one of the first embedding agents used successfully in electron microscopy. They give blocks with good sectioning properties and sections from these blocks produce images with good contrast in the electron microscope without always requiring the use of a heavy metal stain. The main disadvantages of methacrylates lie in the unevenness of polymerization which causes disruption of the ultrastructure, the instability of sections in the electron beam and the shrinkage of the block as it hardens.

The epoxy resins Epikote and Araldite are now more commonly used. Very little shrinkage occurs as the block hardens, and sections from them are stable in the electron beam, though they lack sufficient inherent contrast in the electron microscope and need to be stained with an electron-dense stain.

EPOXY RESINS

The epoxy resin type of embedding media consist of variable proportions of the epoxy resin, hardeners and an accelerator. They are mostly viscous, yellow fluids with a rather pungent smell.

Warning. Skin contact with epoxy resins can cause chronic dermatitis. Adequate precautions, for example wearing rubber gloves and using scrupulously clean techniques, are necessary when handling them.

Epikote (Epon in U.S.A.)

Preparation of resin

Epikote A

Epikote resin 812 (Epon 812 in U.S.A.)	73 g (60 ml)
Dodecenyl succinic anhydride	97 g (100 ml)

Epikote B

Epikote resin 812	121 g (100 ml)
Methyl nadic anhydride	108 g (89 ml)

The mixtures A and B are made separately. Mix thoroughly, preferably using an electric stirrer; be careful not to stir in too much air. Water vapour stirred in from the atmosphere can impair the setting properties of the resin.

Mixing the constituents of the resin by weight in a disposable polyethylene beaker or similar container is more convenient because washing glassware made sticky with residual resin is avoided. The two resins can be kept at 4° C until required for use.

Mixing the resin

Whilst being mixed, the resin gives off an odour. The possible ill effects on health of resin vapour are unknown. Therefore, large volumes should be mixed in a fume cupboard. Before use, allow mixtures A and B to attain room temperature if they have been stored at 4° C. This avoids condensation of atmospheric moisture on the surface of the resin.

The stock mixtures A and B are mixed with the accelerator as follows:

Epikote A	42 g (40 ml)
Epikote B	73 g (60 ml)
DMP 30	1.8 g (1.8 ml)

Weigh the resins and accelerator DMP 30 directly into a polyethylene beaker. Mix thoroughly either stirring by hand for

15-30 min with a glass rod or using a variable speed electric motor. Choose a mixing speed that will not stir air into the mixture. After mixing, the resin is ready for use.

Mixtures of A and B can be used in varying proportions to alter the final consistency of the block. A higher proportion of Epikote A in the mixture will give a softer block; conversely, a higher proportion of Epikote B will give a harder, more brittle block.

The amount of DMP 30 may also be varied. A smaller quantity, for example 1 per cent, rather than the recommended amount 1.8 per cent, gives a softer block; a larger proportion, for example 3 per cent, gives a hard, more brittle block. Blocks which are either very hard or very soft will be difficult to section.

Benzyldimethylamine (BDMA) can be used as an accelerator with Epikote resin. Used in the same proportion as DMP 30, it gives a block of the same hardness with a tendency to be more brittle.

There seems to be no deterioration of the setting qualities of repeatedly frozen and thawed resin for up to a month if it is stored in a freezer at $-20°C$ and thawed before use. After several weeks, resin becomes too viscous to pour properly and consequently does not penetrate tissue very well. At this stage it should be discarded. Do not pour surplus resin down the drains as it may solidify. Excess resin should be placed in a large bottle to be collected with refuse.

Embedding the tissue

The tissue is kept in the same glass vial until the final stage, and the fluids are changed with a Pasteur pipette. The solvents used for epoxy resins tend to dissolve plastic containers.

When processing small blocks of tissue, 5 ml glass vials, size 23 mm × 24 mm, with polyethylene snap-on lids are used.

Embedding schedule

Notes on each stage of the embedding process are given at the end of the schedule.

1. After the tissue has been fixed in osmic acid, rinse the blocks in distilled water for 2 min.
2. Dehydration.

70 per cent ethanol	2 changes of 15 min each.
90 per cent ethanol	2 changes of 15 min each.
100 per cent ethanol	2 changes of 15 min each.

3. Propylene oxide	2 changes of 30 min each.
4. Two volumes of propylene oxide mixed with one volume of resin mixture	60 min.
5. One volume of propylene oxide mixed with two volumes of resin mixture	60 min.
6. Resin mixture	2 h at 37° C.
7. Embed in capsules.	
8. Polymerize	18 h at 60° C.

Notes

Stage 1. Use a Pasteur pipette to replace the osmic acid by distilled water. At all stages of the embedding schedule, discard waste fluids into a glass beaker to catch tissue accidentally carried over with the fluid. Tissues can be lost if fluids are discarded straight into the sink from the embedding vial. To improve the penetration of tissues by the fluids, swirl the blocks around several times in each of the processing solutions.

Stage 2. Ethanol frequently specified for dehydration can sometimes be difficult to obtain owing to problems that can arise over Custom and Excise regulations. Analar isopropanol is a good substitute and can be obtained easily. Tissue can be stored overnight in 70 per cent alcohol if fixation is performed late in the day. The embedding schedule is continued from this stage on the following day.

Stage 3. Propylene oxide is highly inflammable and has a toxic vapour. Appropriate precautions should be taken.

Stages 4 and 5. Make the propylene oxide/resin mixtures in a separate container, not in the vial containing the blocks. This prevents violent agitation of the tissue.

Stage 6. To allow excess propylene oxide to evaporate, remove the lids from the pots at this stage.

Stage 7. Embed the tissue in gelatin capsules type 00 in the following manner: Remove the lid from the base of the capsule. The lid has the wider of the two diameters and will not fit into standard ultramicrotome specimen holders. It has been found that a discarded photographic plate box lid with holes in it makes a convenient holder for gelatin capsules (Fig. 1). These

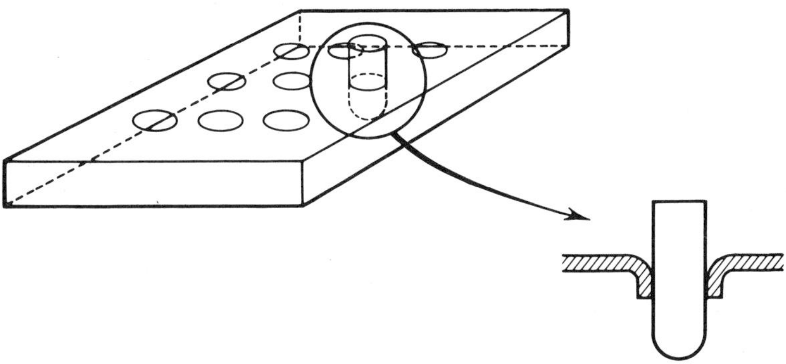

Fig. 1. A photographic plate box lid used as a holder for capsules.

boxes are more convenient than custom made wooden holders as resin is inevitably spilt onto the capsule holder and builds up after several uses.

A system of numbering the individual blocks is important and should be adopted. This is preferable to numbering a box and placing unlabelled blocks in it. This latter system can prove disastrous if several boxes are mixed up. The following system of individual block identification can be successfully employed. Block 'd' of specimen number 80 prepared in 1965 is labelled 65/80 d.

Make the label from white paper using waterproof drawing ink and a mapping pen. (Ball-pen ink dissolves in epoxy resin.) The label is cut so that its length is approximately two-thirds of the circumference of the capsule (Fig. 2). Place the label in the capsule one-third of the way down from the top and press it around the circumference with a pair of forceps. Fill the capsule with Epikote resin mixture. After its final impregnation with resin, drop the tissue onto the surface of the resin in the capsule. The block will slowly sink to the bottom of the capsule before the resin polymerizes. A large number of capsules can easily be filled with resin from a disposable conical polyethylene squeeze bottle.

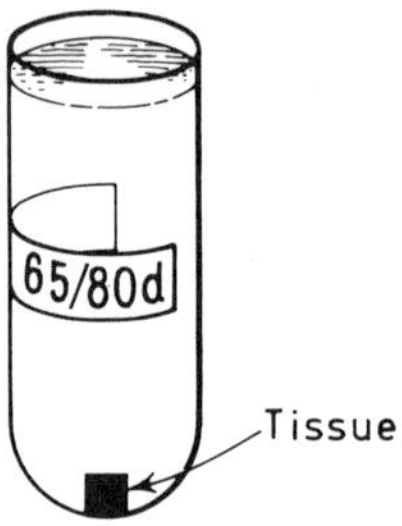

Fig. 2. Embedding capsule complete with identification label showing size and position of the label.

Luft (1961) originally suggested polymerization at 37° C, 45° C and 60° C sequentially for 24 h at each temperature. In practice a single overnight polymerization at 60° C has been found to give blocks of adequate hardness. Blocks that have to be left polymerizing for two or three days are best incubated at 45° C only. This can be useful for weekends.

When polymerized, the blocks should be very hard. Test this by squeezing the label end of the block between the fingers after the resin has cooled to room temperature. Any impression of elasticity indicates that the resin is too soft. There are three possible explanations.

First of all, the block could need further incubation at 60° C, as the incubator may have been below temperature. Once a block has been removed from the incubator and allowed to cool, attempts to increase the hardness by re-incubating it for a longer period or by increasing the temperature may not be successful. The reason for this is obscure, but test capsules filled with resin rather than containing tissue have been found to be useful for testing hardness. The tissue-containing blocks can thus be left in the incubator for a longer period or at a higher temperature without being allowed to cool. Secondly, insufficient accelerator (DMP 30) may have been added to the original Epikote mixture. Thirdly, the resin may not have been stirred sufficiently to ensure a homogeneous mixture.

If the block is firm enough, it is ready for sectioning. The blocks can be stored in small cardboard boxes. We have found that keeping each grid in a separate gelatin capsule after examination in the electron microscope is very useful. Choose a box of such a size that light microscope slides and grids in gelatin capsules can be stored in the same box as the blocks from which they were made.

EMBEDDING MOULDS

Gelatin capsules

Gelatin capsules are available in a range of diameters. Size 00 capsules are normally used for embedding small tissue blocks. These capsules are considerably less expensive than the preformed polyethylene types.

Polyethylene capsules

BEEM capsules and TAAB capsules (Fig. 3) produce resin blocks with the same diameter as size 00 gelatin capsules. The tissue is contained at the apex of a pyramid or cone, which is formed by the mould. These capsules are easily removed from the solid resin after a cut has been made along the full length of the capsule with a razor blade.

As both types of capsule produce blocks with a preshaped end, very little trimming of the block is necessary. This saves time and reduces the amount of resin dust in the laboratory. A snap-on lid attached to both types of capsule can be discarded to save time when a large number of capsules is being used. Whether the capsule is covered by its lid or not is immaterial.

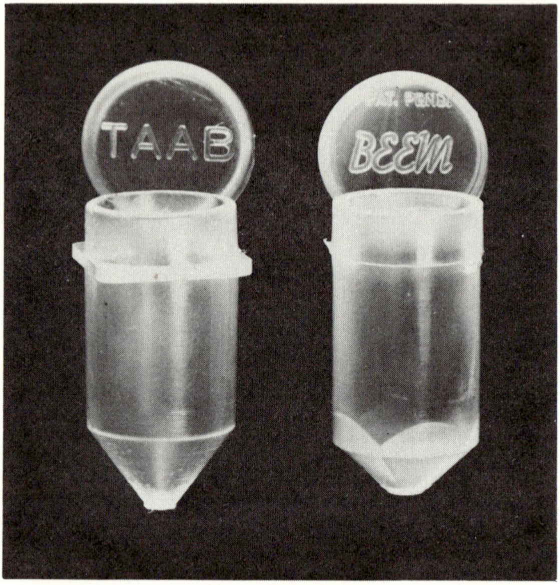

Fig. 3. Two types of polyethylene capsule. TAAB capsules produce conical blocks; BEEM capsules produce pyramidal blocks. The lip near the top of the TAAB capsule prevents it from slipping through capsule holders.

Flat embedding

Aluminium foil can be used to fashion moulds for embedding large, thin specimens. As a precaution against the resin sticking to the aluminium, a very thin film of light oil is spread over the surface with absorbent tissue before the foil is bent to a mould. Cross-sections of very thin pieces of tissue, for example teased nerve fibres and small teased blood vessels, can be obtained using the following technique:

1. Make an aluminium foil mould 10 mm deep, large enough to accommodate the specimen. Fill the mould with resin to a depth of three times the thickness of the specimen. Lay the resin-impregnated tissue lengthwise in the mould (Fig. 4). Polymerize the resin at 60° C overnight.

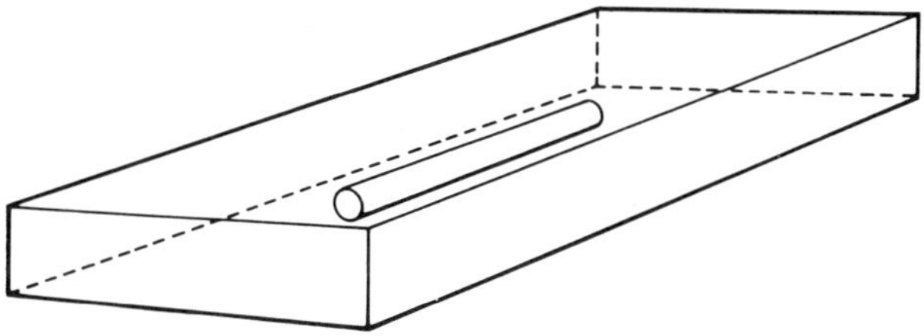

Fig. 4. Aluminium foil mould containing tissue laid lengthwise.

2. With a small hacksaw, cut around the specimen embedded in the resin to form a long, thin block (Fig. 5). This block can then be cut into shorter lengths suitable for clamping into a vice-type ultramicrotome specimen holder. Alternatively, place the block lengthwise in a gelatin capsule and fill with resin which is then polymerized (Fig. 6).

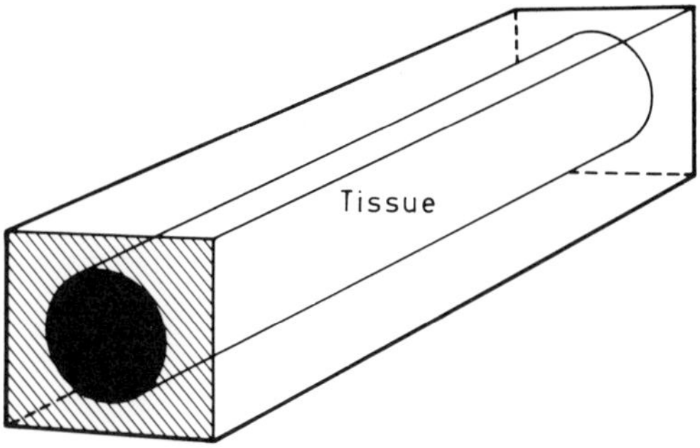

Fig. 5. Block cut with hacksaw from resin in aluminium foil mould.

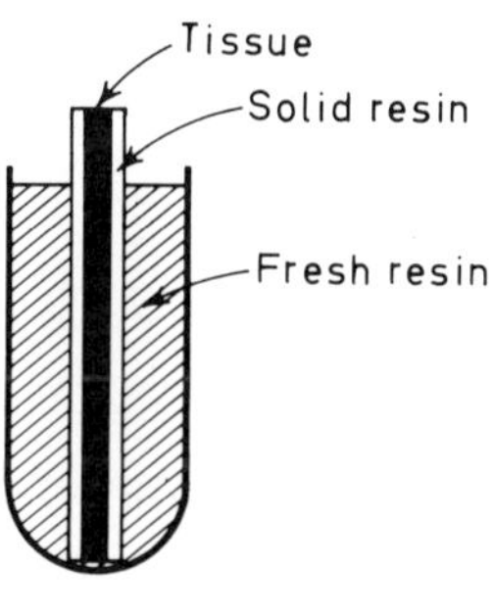

Fig. 6. Block from aluminium foil mould embedded lengthwise in a gelatin capsule to produce cross sections of the specimen.

Araldite

Araldite is an epoxy resin with similar properties to Epikote. The following formula can be used (Glauert and Glauert, 1958).

Araldite CY 212	55 g (50 ml)
Dodecenyl succinic anhydride	49 g (50 ml)
DMP 30	2.5 g (2.5 ml)
Dibutyl phthalate	5 g (5 ml)

Harder blocks are made by using a smaller amount of dibutyl phthalate. Instead of using glassware, weigh the constituents into a disposable polyethylene beaker, and mix thoroughly.

Embedding schedule

1. After the tissue has been fixed in osmic acid, rinse the blocks in distilled water for 2 min.
2. Dehydration.

70 per cent ethanol	2 changes of 15 min each.
90 per cent ethanol	2 changes of 15 min each.
100 per cent ethanol	2 changes of 15 min each.

3. Propylene oxide	2 changes of 30 min each.
4. Two volumes of propylene oxide mixed with one volume Araldite resin mixture	60 min.
5. One volume of propylene oxide mixed with two volumes of Araldite resin mixture	4 h.
6. Araldite resin mixture	12 h.
7. Embed in capsules.	
8. Polymerize	18 h at 60° C.

POLYESTER RESIN

Polyester resins give similar results to epoxy resins. Vestopal W is an example of a polyester resin.

WATER-SOLUBLE RESINS

Tissue is embedded by passing through a graded series of resin/water mixtures with increasing strengths of resin. The normal dehydration and propylene oxide stages are omitted, avoiding extraction of tissue components by these solvents. Durcupan and Aquon are examples of these resins. Aquon is a water-soluble extract from Epikote resin.

EMBEDDING FAULTS IN RESIN BLOCKS

If an attempt is made to achieve reproducible results by following a standard technique, very few problems should occur when using these resins. Blocks that are difficult to section are usually the result of insufficient mixing of the resin (this can give blocks which may be either too hard or too soft), insufficient polymerization time, or too low a polymerization temperature. The dehydration stages rarely cause problems as long as the 100 per cent alcohol is not contaminated with water.

3 Knifemaking

In the early days of electron microscopy, sectioning techniques were not sufficiently advanced to allow sections to be cut thin enough to permit penetration by the electron beam. In order to produce sections of the required thinness (30-80 nm thick) from biological tissues, histological microtomes were individually adapted, and a variety of knives were used. Consistent success was not obtained until the advent of specially designed ultramicrotomes, the use of glass knives, and later the use of diamond knives. A considerable amount of practice is required to achieve a reasonable degree of proficiency in the production of these handmade glass knives. Another major advance was the introduction of a knifemaking machine which made it possible for inexperienced workers immediately to produce usable knives, without having to learn the art of making glass knives by hand.

THE LKB KNIFEMAKER (FIG. 7)

A knifemaker enables the novice who follows the maker's instructions to make excellent knives after a very short period of familiarization.

Suitable glass, already cut into strips of the appropriate width, can be purchased. An instruction manual gives comprehensive instructions on the production of knives with various edge angles (Fig. 8). Knives with a nominal edge angle of 45°, formed by cutting a square piece of glass diagonally, are used to section blocks of normal hardness. Blocks which are too soft to be sectioned with these 45° angle knives can sometimes be sectioned

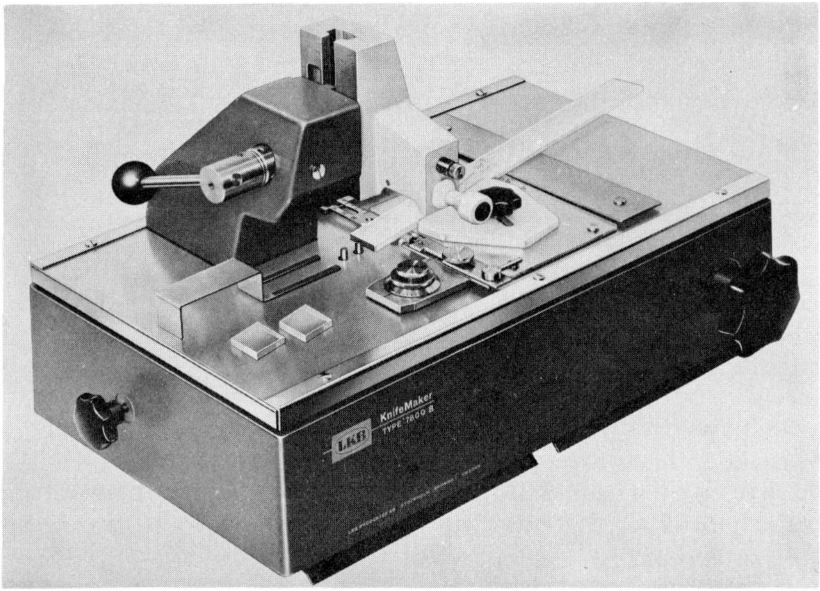

Fig. 7. The LKB knifemaker (photograph by kind permission of LKB Produkter AB).

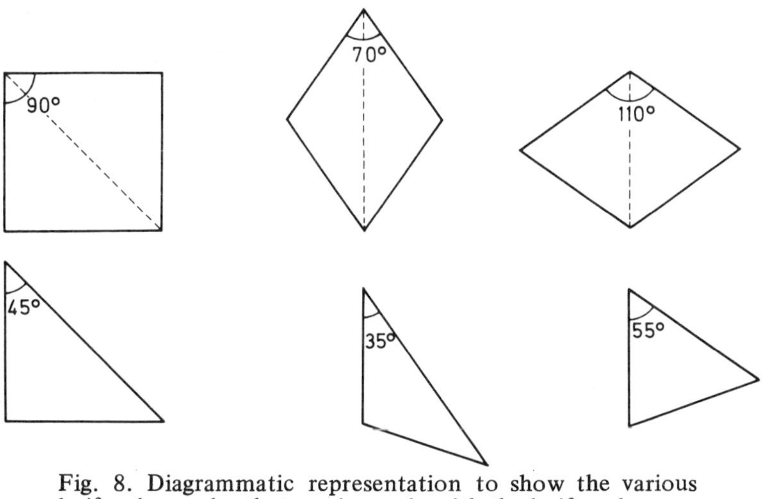

Fig. 8. Diagrammatic representation to show the various knife edge angles that can be made with the knifemaker.

with a knife having an edge angle of 35°, which is cut from a rhomboid-shaped piece of glass.

These knives have a thin, very sharp cutting edge which is useful for cutting soft blocks, but which is easily blunted by harder blocks. These are not recommended for routine use due to the fragility and short life of the cutting edge. To section hard blocks, knives with an edge angle of 55° can be made.

Glass, 6 mm thick, which is normally used in the knifemaker, gives a usable length of cutting edge which is longer than that of knives made by hand where 5 mm thick glass is commonly used. Glass strips of two widths are available for use with the knifemaker.

Glass 25 mm wide

This is used to make knives to fit the LKB, Cambridge Huxley mark 2 and Porter-Blum ultramicrotomes. The knifeholder of the mark 1 Cambridge Huxley instrument will not hold these knives firmly.

Glass 38 mm wide

Knives for the mark 1 Cambridge Huxley ultramicrotome are made from glass of this width.

Laboratories in possession of both LKB and mark 1 Cambridge Huxley ultramicrotomes will be using knives made from glass of both widths. The constant alterations to the settings on the knifemaker is irritating in a busy laboratory. This can be avoided by using the knifeholder adaptor described in the *Journal of Medical Laboratory Technology* vol. 25, pp. 146-147, 1968.

This adaptor, which can easily be made from Perspex, enables knives made from 25 mm glass to be held in the knifeholder of the mark 1 Cambridge Huxley ultramicrotome without any alteration to the instrument itself. These knives are also used in the LKB Ultrotome.

HAND-MADE KNIVES

If a knifemaking instrument is not available, knives can be made by hand with glazier's pliers. With skill and much practice, quite serviceable knives can be fashioned though they tend to vary slightly in size, and inevitably there will be a high reject rate. The basic requirements are:

1. Glass
2. Glass-cutting wheel
3. Glazier's pliers.

Glass

There are numerous types of glass having different physical properties. Some glass is very brittle; other glass tends to be soft. The choice of glass is a compromise between a good cutting edge and the durability of that edge.

Plate glass, 5 mm thick, known to the glass trade as 'drawn' or 'float' glass, is preferred. Soft glass will give a higher percentage of usable knives than a hard glass though cutting edges made from soft glass will not last as long as those made from the harder type. On request, glass merchants can often select a soft or hard glass from their stock.

Glass-cutting wheel

The cutting wheel (which must be of good quality) should rotate freely without wobbling on its spindle. Glass writing diamonds are unsuitable because they normally make a score that is not deep enough to ensure a straight start to the glass break.

Glazier's pliers

The pliers supplied with some ultramicrotomes have been found to be unsatisfactory. Instead either Red Devil pliers, type 210, 8½ in, made in the United States or Elliott Lucas 10-in glazier's pliers, made in England are recommended (Fig. 9).

Some workers claim success with a technique in which the glass is fractured by a pulling motion, utilizing two pairs of similar glazier's pliers. Breaking the glass with a single pair of pliers is preferable. In this method a piece of thin copper wire attached to the lower jaw acts as a fulcrum across which the glass can be broken in a controlled manner.

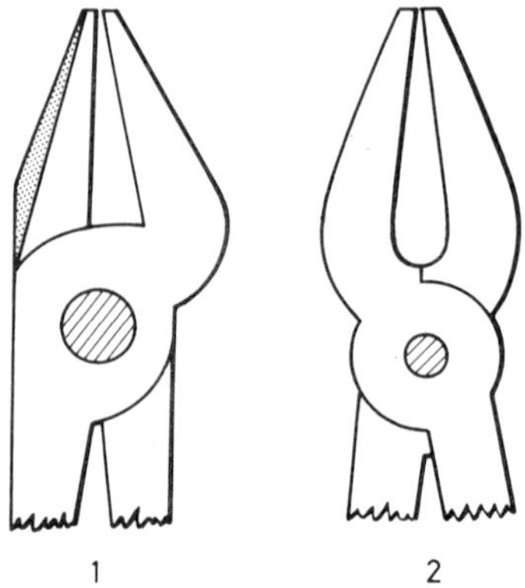

Fig. 9. Jaw shapes of 1. Red Devil pliers, 2. Elliott Lucas pliers.

Preparing the Pliers for Glass Breaking

1. A 40 mm length of copper wire with plastic insulation is positioned longitudinally across the centre of one jaw (Fig. 10a). The wire should be of the single-core type. The Red Devil pliers work best with the wire positioned on the more prominent jaw. Elliott Lucas pliers have symmetrical jaws.

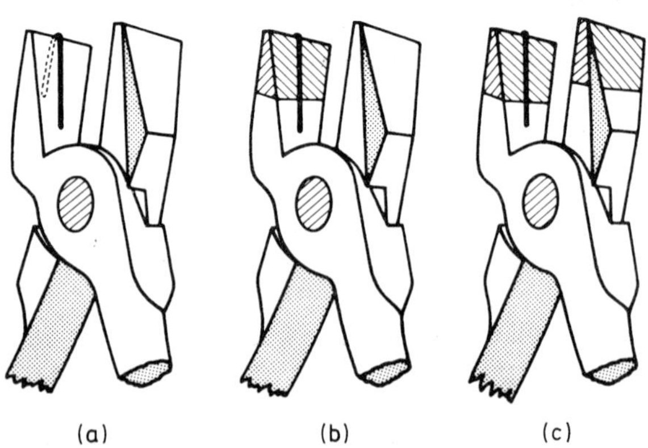

Fig. 10. The preparation of Red Devil pliers for glass breaking.

2. The wire is bound in place with a single layer of 24 mm-wide first-aid plaster which is level with the top of the jaw and which holds the wire central (Fig. 10b). Self-adhesive fabric tape without gauze backing is preferable to the plastic type.
3. Wrap a layer of the same plaster around the other jaw. This acts as a cushion (Fig. 10c). Make sure that it is flush with the top of the jaw. The pliers are now ready for use. Although the copper wire soon becomes flattened in use, there is no adverse effect on the knifemaking properties, which may even be improved. Ultimately, the tape and wire will become so frayed that replacement will appear necessary. There is no point in replacing them as long as good knives are being made.

Procedure

Knives are made by the following technique:
1. Strips are made from a large square piece of glass obtained from a glazier's works.
2. Small squares are made from the strips.
3. Knives are made from the small glass squares.

Making glass strips from a square of glass

1. Take a square piece of plate glass 30 × 30 cm and 5 mm thick. (See note on the selection of glass.) Wash the glass in detergent to remove all traces of grease and dirt.
2. With the cutting wheel, score a line across the centre of the glass from edge to edge to bisect it (Fig. 11). The score should be made with one run of the wheel. If a discontinuous score is made, the glass will tend to break unevenly. Avoid chipping the edge of the glass at the end of the score by running the wheel off the edge of the glass and hitting it with the metal wheel holder. The technique of scoring evenly is an acquired skill. An ample supply of glass should be available for practice.
3. Place the glass squarely in the jaws of the pliers with the score mark directly over the wire fulcrum. By increasing pressure, a crack can be made to extend fully along the score.

Examination of the edge of the glass strip which has been broken by scoring along its full length will show a series of surface

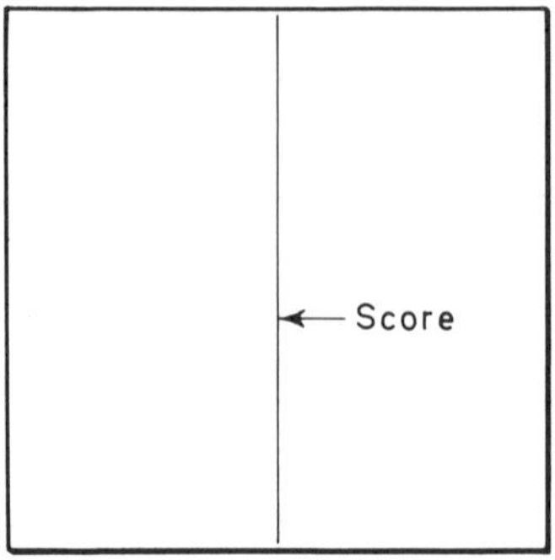

Fig. 11. A 30 × 30 cm piece of glass, which when broken along the central score, will yield two rectangles each 30 × 15 cm.

imperfections and stress lines across the break from the score. In order to overcome the imperfections that may be introduced into the cutting edge of the knife, the free-break technique is employed.

Free-break technique

The free-break technique consists of dividing an area of glass into halves, by making a very short straight score line in order to start a crack which can be made to extend to the other edge.

1. In the middle of one of the long edges of the rectangle of glass obtained by halving the large square of glass, make a short score extending towards the centre (Fig. 12).
2. Place the fulcrum of the glazier's pliers directly under and in line with the score; then, squeeze the glass with increasing pressure until a smooth break starts at the score and continues in the same direction across the glass. The successive glass strips are each halved in the same direction using this technique until pieces 38 mm wide and 150 mm long are obtained (Fig. 13).

Difficulties are encountered when the free break of a 76 mm × 150 mm strip is attempted as the crack sometimes deviates from the centre. To overcome this, score the entire length of the bisecting line to give two strips, one with a free break and a scored break, the other with two scored breaks.

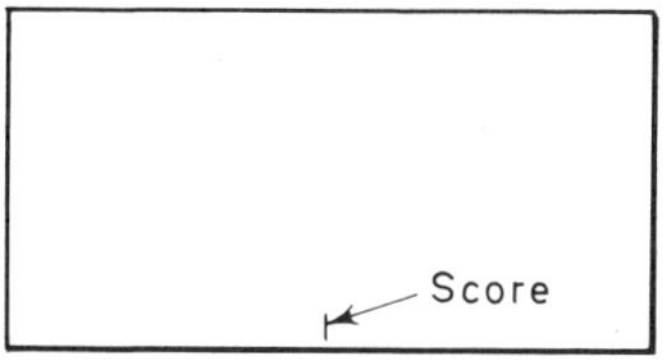

Fig. 12. A 30 x 15 cm rectangle of glass showing the short central score approximately 15 mm long, which is made as a start for the 'free break'.

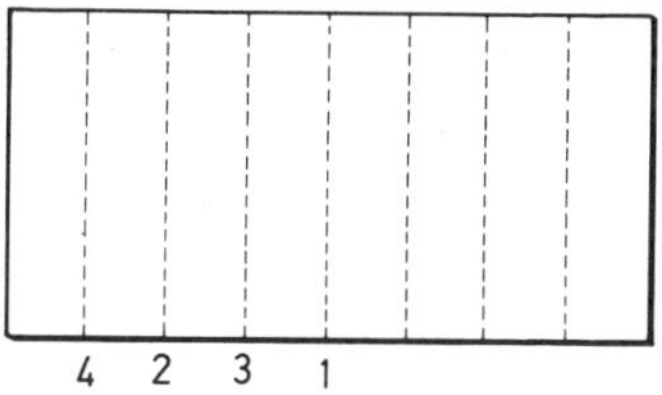

Fig. 13. The sequence of 'free breaks' necessary to make glass strips following break 1, which forms two identical squares from the rectangle.

Making knives from a strip of glass

Knives can be made from the 38 mm x 150 mm strips as follows:

1. Make score 1 (Fig. 14) parallel to the top edge (AB) of the glass strip and of equal distance to the width of the glass from it. The score should start and end approximately 2 mm from the glass edge. Do not touch the edge of the glass with the cutting wheel as this will cause a localized area of damage on the portion of glass from which knives will be made.
2. Make score 2 (Fig. 14). This score is conveniently made using a length of glass as a ruler. The position of the first score can be

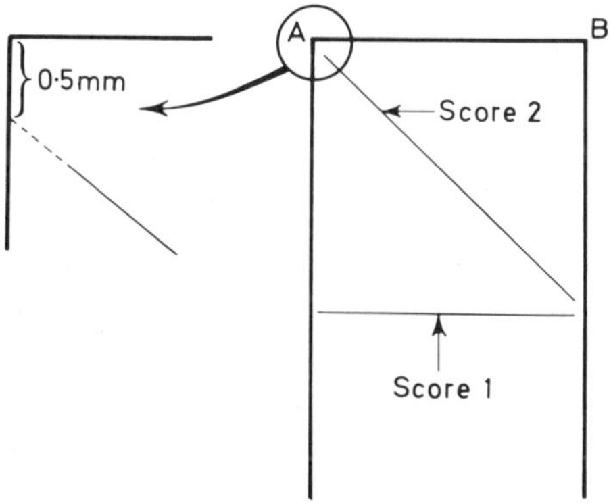

Fig. 14. The position of the two scores needed to make a knife from a glass strip.

seen through the glass enabling accurate positioning of the second score which starts and finishes 2 mm from the glass edges. The 0.5 mm measurement shown in the diagram (Fig. 14) is the distance from the corner of the glass to a point where the score would meet the edge of the glass if continued in a straight line. This distance is critical and is optimally between 0.5 and 1 mm. Within these limits, cutting edges are formed parallel to the straight edge of the corner (Fig. 15). Too

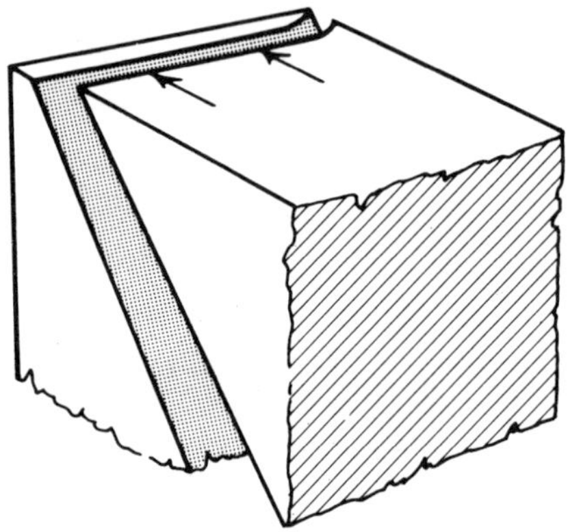

Fig. 15. The arrows indicate the cutting edge, which is formed parallel to the corner of the square of glass.

small a distance from the corner will cause the corner itself to break. A score mark made a large distance from the corner will result in knives with very curved edges. Knives resulting from breaks occurring in either of these positions can sometimes be used for rough block trimming.

The small jig (Fig. 16) supplied with Cambridge Huxley ultramicrotomes is useful as a guide for the diagonal score. When the jig is applied to the top left-hand corner of a glass strip, the distance that the score will be from the corner of the glass is adjusted by a small screw. The top edge of the side arm is used as a guide for the cutting wheel. Once the screw's optimal position is found, it can be locked in order to give a constant starting point and angle for the score.

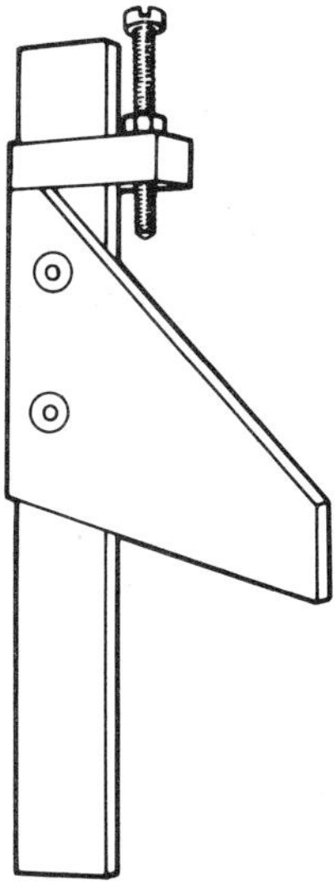

Fig. 16. Jig supplied with the Cambridge Huxley ultramicrotome.

3. Break along score 1 to separate the square from the strip.
4. Grip the square of glass obtained from stage 3 diagonally in the pliers and break it along the score in a slow, controlled manner (Fig. 17). Knives made from slow, controlled breaks are generally superior to those produced with a sudden crack.

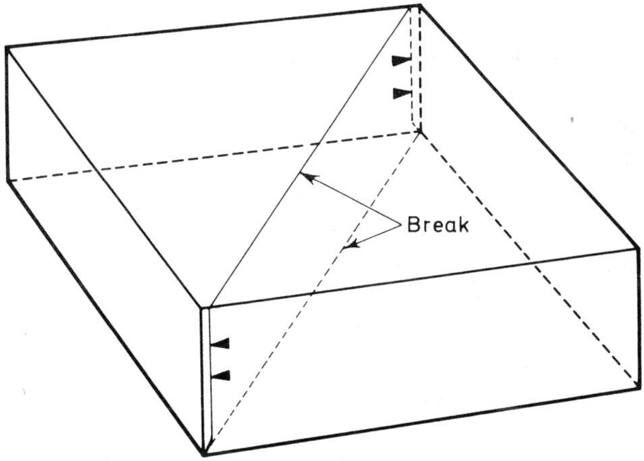

Fig. 17. The arrows indicate the cutting edges formed when a square of glass is broken diagonally.

Because hand-made knives have a base of 38 mm, it is necessary to cut off the lower acute angled corner in order to insert the knife into the mark 1 Cambridge Huxley ultramicrotome knifeholder (Fig. 18a).

For the same reason, the base of the knife must be removed in order to fit the LKB Ultrotome knifeholder. Reduce the height of hand-made glass knives by cutting off a strip approximately 10 mm wide (Fig. 18b). This is done by making a short score mark and using the glazier's pliers.

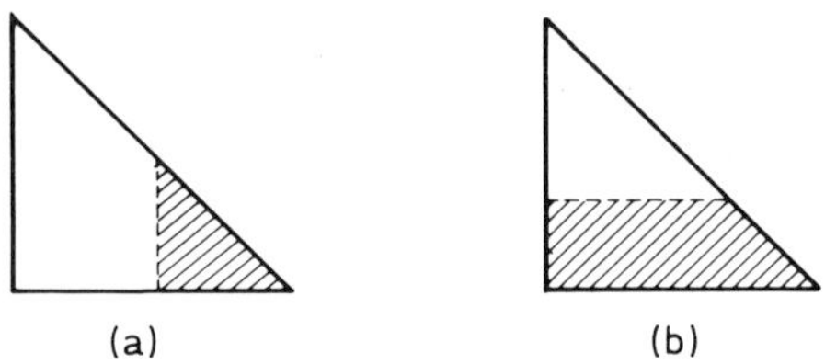

Fig. 18. The shaded portion of handmade knives is removed to fit (a) the Mark 1 Cambridge Huxley ultramicrotome, (b) the LKB Ultrotome.

Theoretically, two cutting edges should be obtained from each square of glass. This is frequently achieved with the LKB knifemaker but rarely occurs when making knives by hand.

Strips of glass with scored edges on each side can be used to make knives provided that the scores are not too deep and the stress marks associated with the scores do not extend to the full width of the edge. This type of strip is placed on the bench with the unscored side of the edge uppermost and facing to the left. Knives are then produced from this edge, with the sharp end of the cutting edge being formed on the unscored aspect. If a Cambridge jig is not available, a piece of 5 mm thick Perspex, 38 mm wide with one end cut at 45° makes a useful guide for the diagonal score.

This method of cutting squares from the end of glass strips and breaking a square diagonally can be applied to glass already cut into long strips supplied by LKB Ltd. This company supplies two types of strip in two widths; one for making knives by hand and the other for use in the knifemaker. Glass for hand-breaking is thinner than that supplied for use with the knifemaker and gives a shorter cutting edge. Both widths of glass (i.e. 25 mm and 38 mm) can be successfully used to make knives by hand.

After the knives have been made, place them, cutting edge uppermost, in a position where they are not liable to be knocked over. A knife that has been dropped or even just knocked onto its side will be sufficiently damaged to make ultrathin sectioning difficult or impossible. A Perspex protective cover placed over the knives keeps dust off the cutting edges. Knives for critical sectioning are best made as required. In practice, deterioration of the cutting edge appears to occur after a few hours exposure to the laboratory environment. Unused knives made one day are probably best used the next day for trimming or for 2 μm sectioning for light microscopy.

Choosing a Good Knife

The best cutting edges are straight, free from edge-blemishes for at least one-third of their length, and have a small spike at one end (Fig. 19). There is always a prominent, curved, smooth stress line

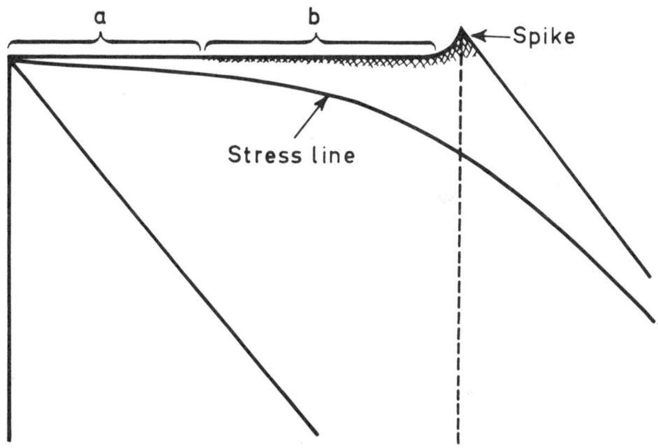

Fig. 19. Diagrammatic representation of a good knife cutting edge. The area 'a' is normally suitable for ultrathin sectioning; area 'b', usually associated with small fracture marks, is only suitable for trimming and 2 μm sectioning.

across the face beginning at the sharp end of the knife (i.e. the end without the spike). This stress line should start right in the corner, not on the cutting edge, and run almost parallel to the cutting edge for its first few millimetres. Minute stress marks always occur at the spike end of the knife; they quickly taper into a smooth edge. The knife edge will show most of its faults in bright relief against the darker glass when examined under the stereo-microscope with an incandescent light source focused on its edge. Scores, stress marks, and nicks in the cutting edge show up brightly. It will be found that only one-third of the knife edge is sufficiently sharp and free from faults to cut ultrathin sections.

The cutting edge of some knives is not straight; varying degrees of upward or downward curve occur. Although knives whose edges curve upwards are very difficult to use for ultrathin sectioning, they can be used for trimming blocks and cutting 2 μm-thick sections for preliminary light microscopy. Of the knives whose edges curve downwards, those with the least amount of curve are usable, but the very curved edges, although sharp, cause the ribbon of ultrathin sections to run into the side of the floating-out trough which makes section collection difficult.

Floating-out Troughs

When ultrathin sections have been cut, it is necessary to pick them up and position them on copper grids for examination in the electron microscope. This is done by attaching a small trough to the knife. The trough is then filled with water on which the sections float as they are cut. Two types of trough are commonly used: individual troughs made from adhesive tape and metal troughs supplied as part of the accessories of the ultramicrotome.

Adhesive-tape Troughs

Wrap a short length of tape around the front of the knife to form a trough level with the cutting edge at right angles to the back of the knife. Any tape projecting from the back of the knife is cut off by running a razor blade along each edge, being careful to avoid touching the cutting edge. The small hole at the bottom of

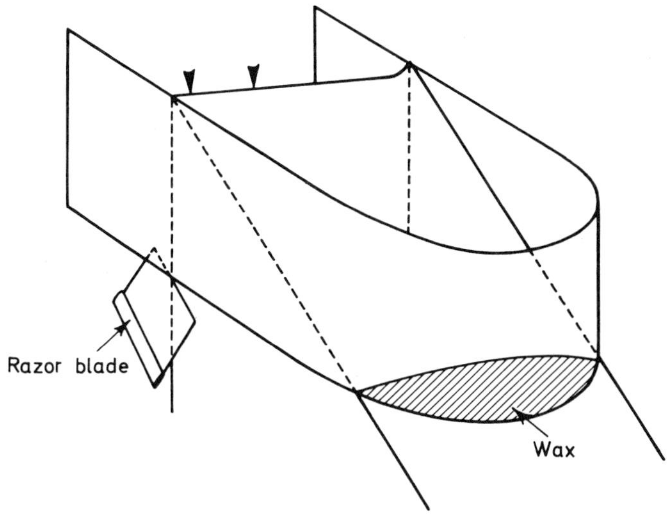

Fig. 20. Adhesive tape applied to a glass knife to form a trough; excess tape is cut off with a razor blade. The cutting edge is marked with arrows.

the trough is filled using molten paraffin wax as a sealant (Fig. 20).

Self-adhesive, waterproof PVC tape, approximately 9 mm wide, is used. The tape is kept in a plastic bag to avoid collecting dust on its edges. This dust will float to the surface and contaminate the sections floating on the water. The dust problem can be overcome if tape 18 mm wide is used. This is cut lengthwise with a razor blade just before use, and the clean cut surface is placed to make the top edge of the trough. These troughs are of course thrown away with the used knife.

Metal Troughs

Metal troughs are attached to glass knives with a few drops of molten paraffin wax (Fig. 21). The edge of the trough is positioned level with the cutting edge of the knife. The working area of these troughs is large. This is an advantage when long, uninterrupted ribbons of sections are required. The problem with

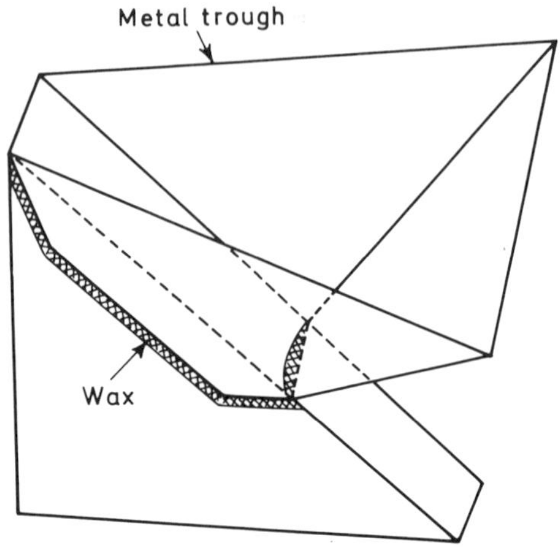

Fig. 21. Metal trough attached to a glass knife showing its edge sealed with paraffin wax.

these troughs is that they have to be thoroughly cleaned after use or storage, and some people find them rather clumsy and inconvenient.

Diamond Knives

Knives made from specially polished diamonds are commercially available. Although these knives are very expensive and easily damaged when carelessly handled, their use is justified when sectioning extremely hard biological specimens that would break the cutting edge of glass knives. The diamond knife is supplied, fixed to a small piece of metal which is inserted into a holder that forms the section collecting trough.

4 Microtomy

In order that biological tissues may be examined in the electron microscope, it is necessary to cut ultrathin sections of the order of 30-80 nm thick. This process is called ultramicrotomy and demands a standardized methodical approach if reproducible results are to be obtained. In this chapter, the various procedures which have to be carried out to produce these sections are described in detail. In brief, the block is sectioned with an ultramicrotome and glass knife to make sections suitable for light microscopy; these sections are then used to choose a smaller area of the block for ultrathin sectioning.

STAGES IN CUTTING ULTRATHIN SECTIONS

1. Trimming blocks–Blocks are trimmed with a razor blade to a shape suitable for 2 μm-thick sectioning.
2. Thick sectioning–An ultramicrotome is used to cut 2 μm-thick sections from the trimmed block. These sections are stained and examined in the light microscope.
3. Trimming blocks for ultramicrotomy–The face of the sectioned block is retrimmed to a size suitable for ultramicrotomy.
4. Ultramicrotomy–Ultrathin sections are cut with an ultramicrotome and collected onto a grid. They are normally stained with solutions of heavy metal salts before examination in the electron microscope.

Trimming Blocks

Blocks formed from gelatin capsules are trimmed into a pyramid with the tissue at the apex. This trimming is not necessary with tissue that has been embedded in BEEM or TAAB polyethylene capsules. It is possible to use an ultramicrotome to fashion the pyramid, by positioning the knife at 45° to the centre axis of the block and by turning the block through 90° as each side is trimmed. Although smooth-sided regular shapes are easily obtained in this manner, the technique is time-consuming. A quicker method using a razor blade is described:

Method

1. Position the block in a small vice, tissue uppermost.
2. Use a small metal file to shape a flat surface on the top of the block and to remove the gelatin and surplus resin almost down to the tissue. A small hand vacuum cleaner with its suction pipe terminating at the vice is useful to remove the large volumes of dust produced.
3. File four flat surfaces each at 45° to the central axis of the block, to form a pyramid, the top edges of which do not quite touch the tissue (Fig. 22).
4. With a small hack-saw, cut off the end of the block opposite to the shaped end. Do not remove the identification label.
5. Remove the file marks from the sides of the pyramid under a

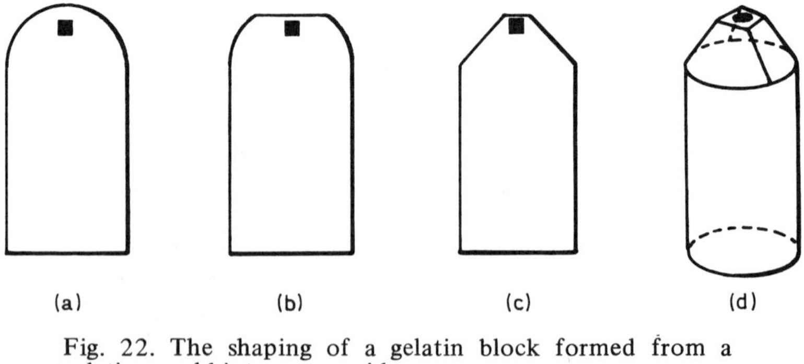

Fig. 22. The shaping of a gelatin block formed from a gelatin mould into a pyramid.

low-power dissecting microscope using a thick one-sided razor blade. The block face should be cut into a square or rectangular shape; trim off all excess resin and leave no ragged edges. The hardness of the resin can be re-assessed at this stage. It should be hard but not brittle or rubbery. An audible high-pitched squeak is frequently heard when cutting resin of the correct consistency. A large area of resin surrounding the tissue will make the flattening of a 2 μm-thick section on a microscope slide very difficult to achieve. The block is now ready to be attached to the microtome for sectioning.

Note. Blocks made from Epikote resin should not be soaked in water to remove the gelatin mould. The resin is hygroscopic and more difficult to section when water has been absorbed. A useful device to support blocks when they are being trimmed under the dissecting microscope consists of a thick lead disc with a central brass spigot or hole depending on the type of specimen holder (Fig. 23).

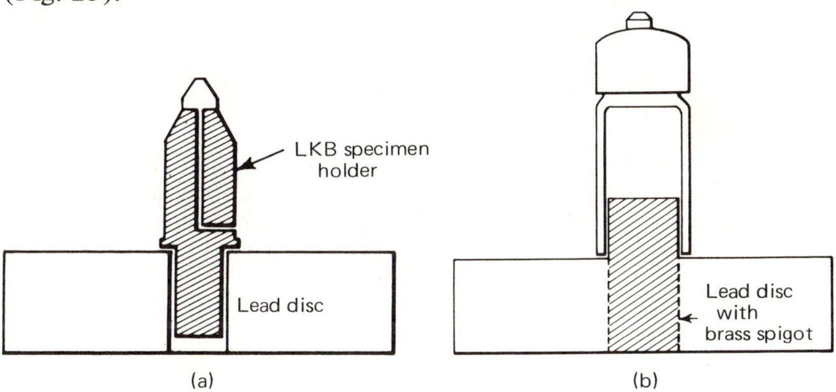

Fig. 23. Hand-made lead discs to hold ultramicrotome specimen holders, (a) for the LKB Ultrotome, (b) for the Cambridge Huxley ultramicrotome.

Ultramicrotomes

Two types of ultramicrotome are commercially available; those using thermal and those using mechanical advance. The thermal advance instruments utilize the expansion of metal when heated to advance the block. Examples of this type of ultramicrotome are the LKB Ultratomes (Fig. 24) and the Reichert ultramicrotome. In the other type of instrument the specimen is advanced by mechanical means. The marks 1 and 2 Cambridge Huxley and Porter-Blum ultramicrotomes are examples of this type of instrument.

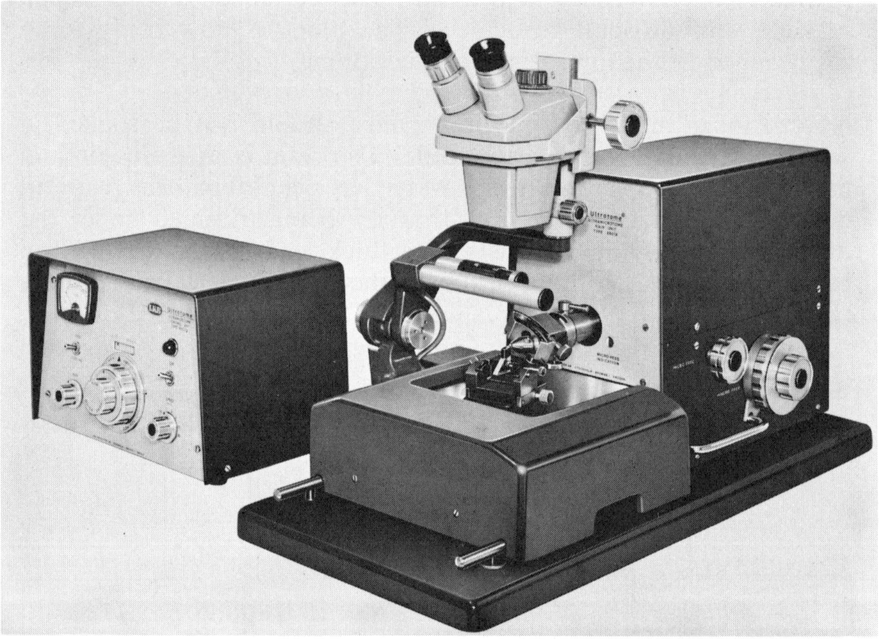

Fig. 24. The LKB Ultrotome III (photograph by kind permission of LKB Produkter AB).

The LKB, Reichert and Porter-Blum ultramicrotomes are automated, cutting consecutive ultrathin sections under the correct operating conditions. Each of these instruments has a range of cutting speeds, good control of section thickness, and there is a considerable range of movement of both specimen holder and knives for alignment of the block face to the knife edge. This versatility of orientation of both blocks and knives can be important. It often allows blocks to be sectioned which would otherwise have to be re-embedded to obtain a section through the required plane, especially if a particular small structure is being studied.

The mark 1 Cambridge Huxley ultramicrotome is much simpler than the automated instruments and is consequently less costly. In use, this instrument can produce excellent ultrathin sections from a great variety of tissues. Control of the cutting speed is achieved by regulation of a screw in the plunger of an oil-filled dash pot. In practice this is not as convenient a control as that used by the automated instruments, but if resin blocks of a standard hardness are routinely produced from one type of tissue this adjustment is not often required. The mark 2 Cambridge Huxley ultramicrotome (Fig. 25), a re-designed version of the existing model, is now

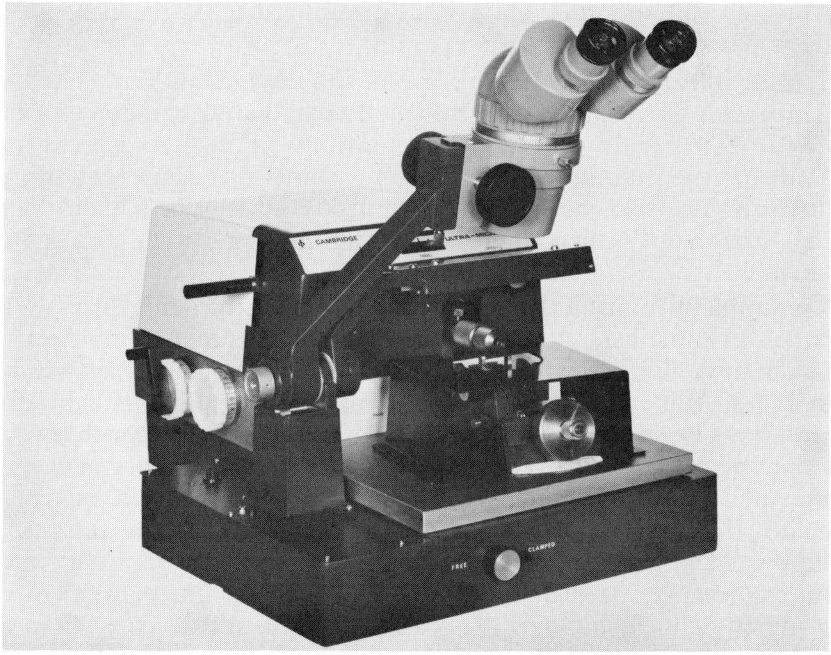

Fig. 25. Mark 2 Cambridge Huxley Ultramicrotome (photograph by kind permission of Cambridge Scientific Instruments Ltd.).

available. It has controls that appear to be easier to adjust than those of the mark 1, and the knifeholder will take knives made by the LKB knifemaker.

An electron microscope unit which is called upon to examine a wide range of tissues should consider the automated ultramicrotome to be an essential part of its sectioning equipment. They are versatile and they enable most sectioning problems to be overcome. The less versatile instruments can be used to section blocks of known consistency. However, the skill of the ultramicrotomist, which is only acquired after much practice and experience, is probably of greater importance than the design features of any specific model of ultramicrotome.

Site of ultramicrotome

When considering the site for an ultramicrotome, the following three points should be borne in mind.

Temperature. The sectioning room should have a fairly stable temperature. Expansion or contraction of metal components in these instruments by rapid fluctuations of a few degrees in ambient temperature is enough to make ultrathin sectioning difficult. Thermal movement thus introduced may be greater than the instrument's own ability to advance to or retract from the block.

Vibration. Vibration transmitted to the instrument through a bench or table can cause variations in ultrathin section thickness, or chatter. This is seen as a series of alternate thin and thick bands in the section, parallel to the knife edge. If the floor is liable to vibration, position the instrument on a substantial bench away from the centre of the room with blocks of foam rubber between the ultramicrotome and the bench. Specially designed benches made from slabs of reinforced concrete faced with terrazzo, supported at each end by a cemented brick pier with a concrete foundation, are used in one laboratory. These can be conveniently made by a local builder. Persistent vibration such as that from a busy road can make ultrathin sectioning impossible unless a vibration proof sectioning table of the type marketed by LKB Ltd. is purchased. These tables are expensive, but they may be necessary in certain circumstances. In many instances, the elimination of vibration is a matter of trial and error and common sense.

Lighting. As it is necessary to see small objects clearly, a high level of lighting must be available. However, it is not a good idea to site the instrument near a window as direct sunlight striking the ultramicrotome upsets its thermal stability and also makes it difficult to see ultrathin sections. A room that is free from distractions and draughts, and which is not used as a general thoroughfare is essential in order to obtain good, consistent results. Keeping the area surrounding the ultramicrotome clean and dust free is also important. However, the most important requisites for the ultramicrotomist are patience and a high degree of concentration.

Thick Sectioning for Light Microscopy

A 1-2 μm-thick section mounted on a glass microscope slide is either stained or examined by phase-contrast microscopy. The sections are used to select suitable areas for ultrathin sectioning for electron microscopy. They are also used because of the excellent cytological detail visible with specially fixed tissue; these sections are thinner than those produced from paraffin wax blocks. The light microscope may yield information which might not be revealed when using conventional methods of embedding and sectioning for routine histological purposes.

The general technique described for thick sectioning is applicable to all ultramicrotomes. However, reference must be made to the manufacturer's operating manual for specific instructions regarding the position, setting and manipulation of the various controls of the instrument in use.

Method

1. Before using an ultramicrotome, ensure that the micro-advance for the specimen is at the start of its travel and not near the end.
2. Mount the trimmed block securely in the ultramicrotome

specimen holder, with the pyramid and 1-2 mm of block projecting from the holder. This is then clamped to the specimen arm of the ultramicrotome.

3. Move the knife towards you, allowing ample room for the knife to be inserted into its holder without touching the surface of the block. The knife should be held securely, with its back edge flat against its holder in order that the correct clearance angle (Fig. 26) can be obtained. The clearance angle is set by tilting a

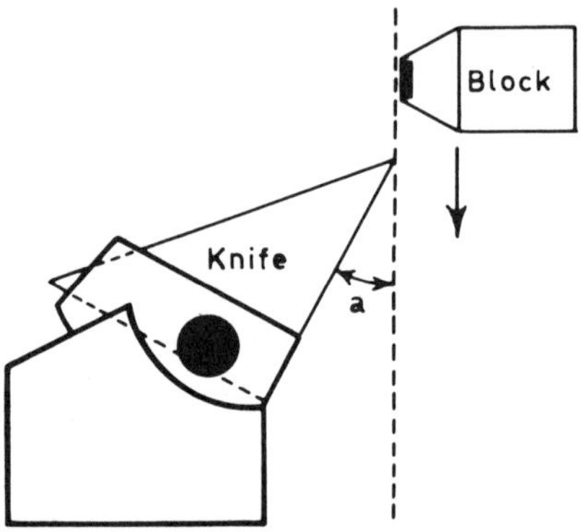

Fig. 26. Showing the clearance angle 'a' between the back of the knife and the plane of the block face.

movable part of the knifeholder to an angle measured by a series of marked graduations. This angle varies with the type of ultramicrotome in use; for example, the Cambridge Huxley ultramicrotome normally sections well with a clearance angle of 7°. The recommended clearance angle setting for the LKB Ultrotome is 4°. In practice, these settings can be altered by one or two degrees on each side of the recommended angle, in order to allow for individual microtomes and blocks of differing consistencies. With an insufficient clearance angle, the block will rub against the back of the knife and fail to section. An angle that is too large causes chatter and missed sections.

4. Looking towards the side of the instrument, advance the knife towards the block until it is approximately 1 mm away from

the trimmed face. Adjust the level of the block until the trimmed edge of the pyramid is 0.5 mm above the knife edge. Arrange the fluorescent lamp to illuminate the block face and knife edge when viewed with the stereomicroscope. Move the knife sideways until the block is positioned over a length of the less sharp end, i.e. the spike end of the knife edge. Loosen the block holder and rotate it on the specimen arm until the lower edge of the block is parallel to the knife edge. Then securely clamp the specimen holder to the specimen arm.

5. Slowly advance the knife towards the block, simultaneously moving the block face up and down until a first contact is made between the knife and the block. At this stage there is no need to perform a complete cutting cycle, as movement upwards to bring the lower edge of the block above the cutting edge and downwards until the top of the block disappears below it are sufficient and less time consuming. Continue this action, making 2 μm advances of the knife between sections until a complete section of the tissue is obtained. Remember to advance the knife only when the whole block is above the cutting edge; this is important. Occasionally remove the debris accumulated on the knife with a pair of fine forceps. Avoid touching the knife edge. It is normally easy to see when a complete section is being cut by comparing the relative amount of resin and dark-coloured tissue in the pieces trimmed from the block face.

   If the trimming process is carried out too quickly by cutting sections from the block which are much thicker than 2 μm, there may be a tendency for small pits to appear in the face of the block. Subsequent sections from that face are liable to develop holes. These holes can only be removed by cautiously trimming the block in steps of less than one micron. Imperfections in the knife edge are also introduced by trimming too thickly. The blunted, chipped cutting edge can cause either a deep score in the face of the block or a ridge protruding from the face. Both can only be removed by very delicate trimming.

6. When the required depth in the tissue has been reached, the block face should be free of scores and holes and should appear

to be polished. Suspend the block above the knife edge. Retract the knife several millimetres from the block and remove the trimmings from the face of the knife with a pair of fine forceps or a very fine brush.

7. Move the knife sideways with the adjustment on the knifeholder until the unused portion of the knife at the sharpest end of the cutting edge is directly below the block face.
8. The cutting edge of a glass knife is rarely absolutely straight over its whole length, and thus the block face has to be matched again to the new edge whenever the knife is moved laterally. This is done by advancing the knife towards the block, moving the block up and down until initial contact is established, and then cutting 1-2 μm sections until the whole face of the block is being sectioned. As soon as the block face and knife edge are fully matched, bring the specimen arm to rest and remove the trimmings from the knife.
9. Select a section thickness of 1-2 μm on the microtome.
10. Cut the first section at a slow speed. If the speed of the cutting action is too fast, the sections tend to curl up into a roll which is difficult to straighten. At the correct speed, 2 μm-thick sections will be flat and will slide down the face of the knife as they are cut. They adhere to the knife only at the cutting edge when the top face of the block is reached. Sections at 1 μm or thinner do not normally slide down the knife as well as thicker ones as they tend to form corrugations. These sections can be flattened when placed onto a drop of water on a microscope slide.
11. One drop of distilled water is placed in the centre of a clean, grease-free glass slide. Use a pair of watchmaker's forceps (No. 4 or 5, with long, straight points) to remove the section from the knife edge. Hold the section by one corner, being careful not to push the points of the forceps through the section or grip it in the middle.
12. Place the section on the surface of the water on the microscope slide where it will float but not flatten out. Flatten the section either by breathing on it until a film of moisture appears on the section and slide or by holding a

piece of cotton wool moistened with chloroform a few millimetres above the section. The breathing technique works well and is convenient.

13. The preparation is dried in an incubator at 60° C. For examination by phase-contrast light microscopy, the section is mounted in a synthetic mounting medium, for example DPX, and covered with a coverglass.

Staining Sections for Light Microscopy

Unfortunately few of the stains routinely used in conventional histology have the ability to colour tissues embedded in synthetic resins. Reasons given for the poor uptake of these dyes suggest that they might not penetrate the resin. It is also possible that the osmic acid alters the staining properties of tissue. Furthermore, the sections are very thin with a very small mass of tissue contained therein, and the amount of stain taken up may be too small to be seen.

Some dyes, however, do appear to stain tissues embedded in resin, and amongst these, toluidine blue is possibly the most widely used.

Toluidine blue stain

Borax	1 g
Toluidine blue powder	1 g
Distilled water	100 ml

Dissolve the borax in the distilled water; warm the water if it is necessary, and follow with the addition of the dye. When the dye powder has dissolved, filter the stain through a Whatman No. 1 filter paper into a clean bottle.

Method. Cover the section with four or five drops of the stain and warm on a hotplate or over a small Bunsen burner flame until steam appears. When using the burner to warm the slide, avoid heating directly underneath the section as this tends to detach it from the glass. Observe the intensity of tissue coloration by tilting the slide until the section is visible. The tissue appears dark blue when correctly stained with no stain in the resin. Allow the stain to cover the section again if it has not reached the required intensity of colour.

When the staining is adequate, wash the slide with distilled water, and allow it to dry at 60° C. Place a drop of synthetic mounting medium on the section and cover with a coverglass which is lowered gently to avoid introducing air bubbles into the mounting medium.

Result. This single stain produces various shades of blue, making identification of tissue components such as nuclei, membranes, and some intra-cytoplasmic organelles easy. Essentially, the stain only provides different degrees of contrast, and specific structures or tissues are not selectively stained.

Cresyl violet stain (Griffin)

Borax	2 g
Cresyl violet powder	2 g
Distilled water	50 ml

Dissolve the borax in the distilled water by heating it to 60° C, and add the cresyl violet powder. A heavy brown precipitate forms. Remove the source of the heat, and add 50 ml of isopropanol in order to produce 100 ml of bright red solution.

Method. When cool, filter this solution onto a resin section on a microscope slide and allow it to remain until the section stains red. Wash the slide in isopropanol and allow it to dry. The mounted section can be examined when it is stained with the cresyl violet only, or a blue nuclear counterstain can be applied as follows.

Wash off the stain with isopropanol. When the slide is dry, place onto the section one drop of a freshly prepared mixture of equal parts of aqueous 0.5 per cent methylene blue and aqueous 0.5 per cent sodium hydroxide. Apply a coverglass to the preparation and observe the intensity of the stain under a microscope using a low-power objective. As soon as the cell nuclei have stained blue and before further staining of the section occurs, wash the stain off with distilled water. Dry the slide at 60° C and mount in a synthetic mounting medium.

Result. This combination of stains produces a section with blue nuclei, and the remaining tissue components stain red, with the mitochondria prominently stained. The cresyl violet stain is less permanent than toluidine blue as the red colour tends to fade to light violet.

Trimming Blocks for Ultramicrotomy

Following examination in the light microscope, an area of the thick section is selected for further examination in the electron microscope. For this, the block has to be retrimmed to produce a smaller face than that from which the 2 μm sections were taken as this face is usually too big to be ultrathin sectioned. Mark the top of the block with ink or a wax pencil as a guide for replacing it in the microtome.

1. Remove the block-holder from the microtome and place it in a suitable holder under the dissecting microscope. It is useful to have a dissecting microscope of the zoom type or one with a triple nose-piece to change magnifications which allow the examination of the block face at a high power. Lower magnifications are used when trimming the pyramid with a razor blade.

   A dissecting microscope is not required when using the LKB Ultrotome. The instrument's own microscope can be positioned vertically, away from the knifeholder, above a clamp attached to the microtome to hold the block firmly. This clamp is available as an accessory.

   It is useful to have the dissecting microscope near the light microscope to compare the surface of the block with the thick section when selecting an area for further examination.
2. Move the source of illumination around the block until a point is reached when detail on its face is apparent. At this stage it is possible to compare the face of the block with the thick section. The area to be selected from the thick section is marked on the face of the block by lightly scoring with a razor blade. An area slightly larger than the intended face is initially marked out to allow for trimming.

   The size of a block that can be ultrathin sectioned depends on the consistency of the resin, the type of tissue, the quality of the knife and the skill of the microtomist. In view of this, the novice is advised to start with very small block faces as they are usually easier to section.

3. Trim the block into a smaller pyramid, keeping the black ink mark as a guide to the top of the block. Remove thin slices from the sides of the block with a clean, sharp razor blade. The shape of the final pyramid is important if good sections are to be obtained (Fig. 27). The sides of the block should consist of

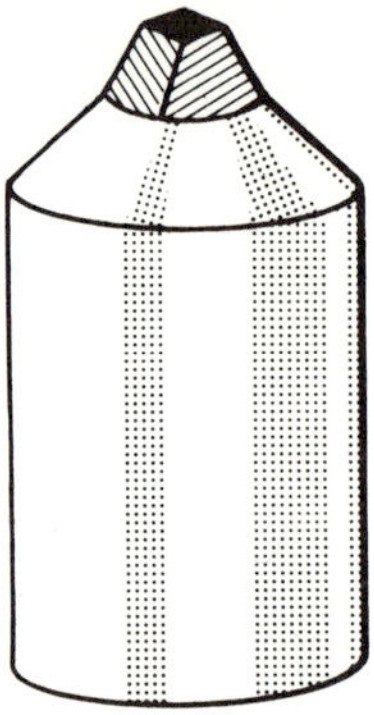

Fig. 27. Ideally shaped pyramid for ultrathin sectioning.

tissue only as surplus resin surrounding the tissue will produce sections with chatter and wrinkles which tend to originate at the tissue resin junction.

The sides of the pyramid should not be steep but at an angle of approximately 45° to give a broad based block with smooth sides. Viewed from the top, the block should resemble that shown in Fig. 28, with the sides sloping slightly inwards, towards the top edge. The lower edge is usually orientated parallel to the cutting edge of the glass knife. The top edge should be not quite parallel to the lower edge but tilted by a very few degrees towards the right-hand side of the block. This ensures that the ribbon of sections will float towards the centre of the trough and not into its left-hand side (Fig. 29), where a 'log-jam' of floating sections would form.

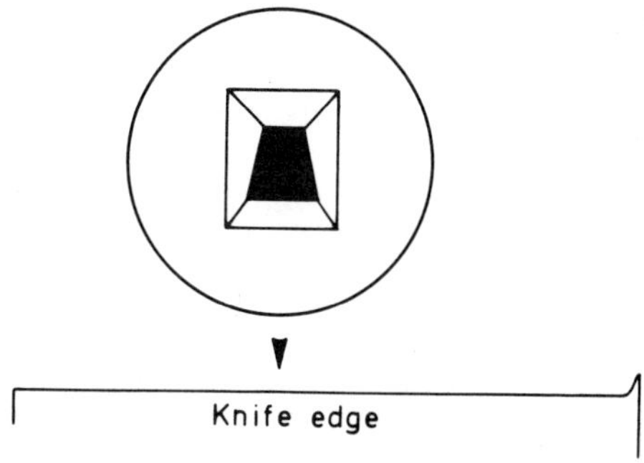

Fig. 28. Plan view of a well shaped block face, showing the orientation of the face to the knife edge.

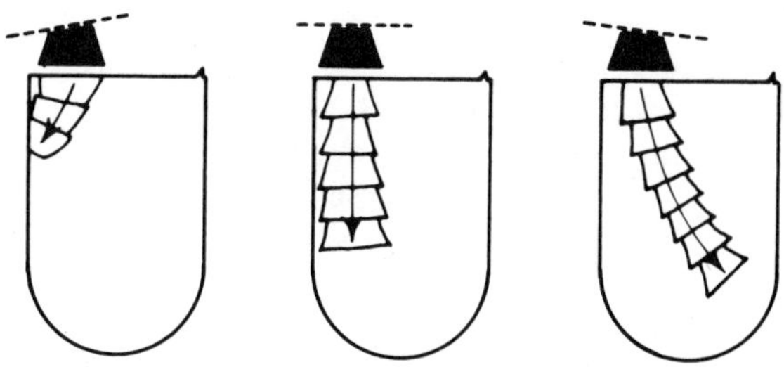

Fig. 29. This diagram shows the effect that the shape of the upper edge of the block has upon the direction of the ribbon of sections on the surface of the water in the trough.

Ultrathin Sectioning

In ultrathin sectioning the block is retrimmed after selecting a suitable smaller area from the thick sections in the light microscope. The block is replaced in the ultramicrotome, and a new knife complete with floating-out trough, is inserted. A portion of the new cutting edge is matched to the block face before being moved laterally to be rematched to the definitive cutting edge, and ultrathin sections which will float on the surface of the water are cut.

1. Replace the specimen holder containing the trimmed block in the microtome, ensuring that it is as near as possible to its original position. At this stage loosely clamp the holder to the

specimen arm as the block will have to be lined up with the knife edge.

2. Position the micro-advance mechanism of the microtome near the start of its travel. It is better done at this stage than when half a ribbon of ultrathin sections has been cut.
3. Retract the knifeholder from the block in order to allow adequate space to insert the knife without touching the block face.
4. Place a new knife, complete with floating-out trough, in the knifeholder, and clamp it firmly at the manufacturer's recommended height. Ascertain that the back of the knife is parallel to the rear face of the knifeholder and is giving the correct clearance angle set on the holder.
5. Position the block with its lower edge 1 mm above the cutting edge of the knife, and parallel to the knife edge. Tighten the specimen holder to the specimen arm.
6. Critically examine the knife edge, using an incandescent light source focused on the edge. (See notes on the selection of glass knives, Chapter 3.) If the knife edge appears to be satisfactory, switch off the incandescent lamp and switch on the fluorescent light.
7. Slowly move the knife forwards until the knife edge and block face are both in the field of view of the microscope but not yet in contact with each other.

   The angle of the microscope in relation to the edge of the knife and block face is now set. This angle is a matter of individual preference. One should be able to see all but the very smallest gap between the knife edge and block and at the same time be able to obtain a good view of the surface of the water filling the trough.
8. Move the knife until the block is adjacent to the portion of the cutting edge to be used later for ultrathin sectioning (Fig. 30a).
9. Cut one or two thick sections (1-2 μm) in order to match the knife to the block face. Remove the debris.
10. Retract the knife edge until there is an obvious gap between it and the block face.
11. Position the block over the part of the edge selected as best for

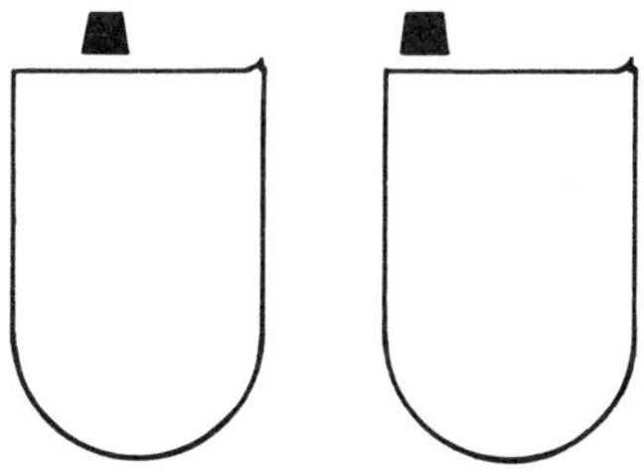

Fig. 30. (a) The block is matched to an area adjacent to the portion of knife chosen for ultrathin sectioning; (b) the block is moved laterally and ultrathin sections are cut.

ultrathin sectioning (Fig. 30b). The block may have to be raised a little if the cutting edge slopes up towards the corner.

12. Use the micro-advance control to match the block face to the new cutting edge. Perform a complete cutting cycle, only advancing the block when it is above the knife edge. The use of the micro-advance is recommended to avoid cutting a thick section on this portion of the edge. This would make it too blunt for good ultrathin sections to be produced. Cut a minimum number of sections less than one micron thick to produce a highly polished block face. When this is achieved, stop trimming and support the block above the knife edge. Remove the debris.
13. A syringe fitted with an angled needle is used to fill the floating-out trough with distilled water to form a meniscus just above the level of the trough.
14. Look through the microscope at the water surface close to the cutting edge, and slowly withdraw water from the trough with the syringe until the meniscus is seen to be below and tapering up to the cutting edge. The water should only just touch the cutting edge. If the water is higher than the cutting edge, it will wet the block face and make sectioning very difficult. Sections will fail to float out and pile up if the water is not actually touching the edge.

Obtain a brilliant white reflection of the fluorescent lamp on the surface of the water in the region of the cutting edge by positioning the lamp by trial and error. This reflection must be present if the interference colours which are used to gauge the thickness of the sections are to be seen. On some occasions, depending on the angle formed by the top of the trough with the back of the knife, the block may have to be positioned at a level below the knife edge and the water level and the position of the lamp adjusted in order to obtain the correct reflection.

15. With the ultramicrotome set to produce the thickest ultrathin sections, start the cutting cycle either automatically or manually. It will be found that most tissues section well at a rate of descent of two millimetres per second.

    The cutting speed regulator of the Cambridge Huxley ultramicrotome is a screw in the damper piston; this screw is adjusted by trial and error until a suitable cutting speed is obtained. Smooth rhythmical cycling is the secret of success with this instrument. Lift the arm at the same speed each time to avoid cutting sections of uneven thickness, taking care not to jar the ultramicrotome at any stage.

    The first two or three sections will be thick and often uneven. These are removed with a pair of fine forceps while the instrument is in the non-cutting phase of its cycle. Great delicacy is required at this stage so that touching either the instrument or the knife edge is avoided. Gradually reduce the thickness of the sections at a stage in the cycle when the block is below the edge of the knife. Avoid making sudden large alterations in section thickness. Unsuitable sections should be removed with fine forceps until sections of the required thickness are produced in a ribbon.

    The interference colours of the sections as they lie on the water are used as a guide to their thickness.

Group A	Group B	Group C
Violet	Medium Yellow	Light Grey
Red	Pale Yellow	Silver
Blue	White	
Green	Dark Grey	
Brown		
Dark yellow		

Group A sections are very thick and are normally unsuitable for electron microscopy.

Group B sections give electron micrographs of good contrast at low magnification, being especially useful for making montages of large areas. These sections are too thick for high resolution work.

Group C sections are excellent for high resolution microscopy at high magnifications but often lack the contrast necessary for low magnification studies.

16. Allow the microtome to cut enough sections to form a ribbon whose length is approximately twice the diameter of the grid. Stop sectioning with the block at rest below the knife. A more evenly illuminated field of view is normally obtained with the block resting in this position.
17. Detach the ribbon from the knife edge by making one quick cutting cycle with the specimen advance set to zero. If this action is carried out correctly, the ribbon will float into the middle of the trough without another section being cut. With the syringe add a very small quantity of water to the trough to make the surface fractionally below the level of the walls. The ribbon can be broken into two with a pair of clean, grease-free, fine-pointed forceps. Hold the forceps closed, and lightly apply the points to the junction of two sections. Allow the forceps to spring open, dividing the ribbon along the join. With practice this can be achieved without the sections attaching themselves to the forceps. The ribbons can be manoeuvred alongside one another by lightly touching the surface of the water with the forceps. The sections are now ready to be picked up onto grids.

Collection of Sections onto Grids

Due to their stability in the electron beam, sections from tissue embedded in epoxy resins can usually be collected onto grids without plastic or carbon support films. The detailed production of support films is given in Chapter 8.

The diameter of the grid upon which sections will be picked up is determined by the type of specimen holder supplied with the electron microscope. Manufactured grids, made from copper, nickel or gold, with a wide range of mesh pattern and size are commercially available. For routine work the cheaper copper grids are the most satisfactory. Nickel grids are desirable when sections have to be immersed in solutions that react with copper to produce a precipitate that may contaminate the sections. For example, the developer used in electron microscope autoradiography attacks copper in a very short time. Grids normally arrive from the supplier ready for use packed in clean polyethylene tubes. They are probably best left in these containers until they are required; remove one at a time with a pair of fine-pointed watchmaker's forceps which are used exclusively for this task. For routine purposes three sizes of grid mesh (small, medium and large) should be kept close to the ultramicrotome to allow a grid suitable for the size of section to be chosen.

Method

1. Pick up a grid by its edge with the watchmaker's forceps. It is useful to place a few turns of thin wire or a small neoprene 'O' ring around the stem of the forceps. This enables the points of the forceps to be kept closed and permits the grid to be manipulated with greater dexterity.
2. Strong surface tension forces seem to be produced when grids are submerged under water prior to picking up the sections. This makes the collection of sections tedious as they move away from the grid as it breaks the water surface. This can be avoided: dip the grid into a 1 per cent solution of Teepol or other surface wetting agent in distilled water, rinse in distilled water, and pick up the sections before the grid dries. Alternatively wash the grid in propylene oxide or carbon tetrachloride, and allow it to dry before use.
3. Submerge the treated grid, edge first, in the collecting trough, disturbing the water surface as little as possible.
4. Raise the grid slowly at an angle of 45° to the surface of the water, allowing the end of the ribbon to adhere to its edge. Withdraw the grid from the water maintaining this angle. The sections should lie across the middle of the grid (Fig. 31).

   Isolated sections can be manoeuvred as close together as possible and then picked up using either of two methods. In

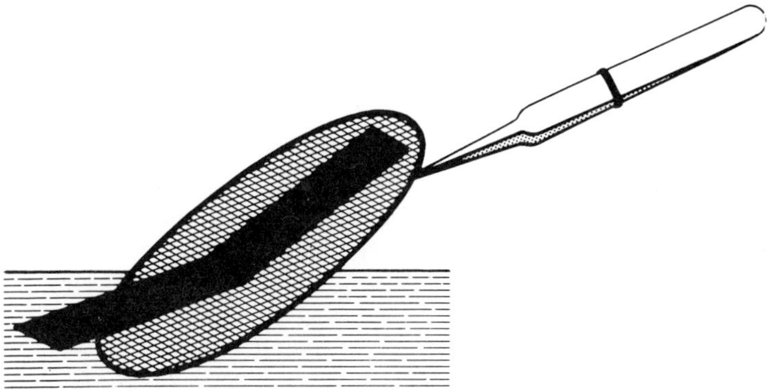

Fig. 31. A ribbon of ultrathin sections adhering to the grid as it is withdrawn at an angle of 45° from the water in the trough.

the first, lift a grid parallel with the surface of the water under the sections, slowly allowing the sections to settle onto the grid as it breaks the surface of the water. It is essential that the grid is pretreated with a surface wetting agent if the sections are to remain in position over the grid as it breaks the surface. In the second method, drop a grid, treated with organic solvent and dried by warming gently, onto the sections. This is a difficult technique and is only used as a last resort.

It is of little importance whether the smooth or matt side of the grid is used. Sections adhere to either side equally well except when using the latter technique described above, where the smooth side of the grid gives a higher rate of success.

5. Keeping the grid in the forceps, remove surplus water from between the points by means of a clean filter paper. Remove water from the grid itself in a similar manner, by touching the edge of the grid with filter paper. This sequence is carried out in this order to prevent any debris between the points of the forceps being carried across the grid by the flow of water and being deposited on the sections. The grid, still clamped in the forceps, is placed in a dust free area and not released until completely dry.

The section is now ready for staining and should be stored on a clean filter paper in a closed Petri dish or a similar enclosed receptacle.

TROUBLESHOOTING

Difficulties in ultrathin sectioning do arise and can be accounted for. The block may be badly matched to the sharpest part of the knife edge, or the level of water in the floating-out trough may be incorrect.

Failure to section

1. Loose knife or block. Check that the block, specimen holder and knife are all firmly held in their correct positions.
2. Soft blocks. A very soft block, due to either poor resin or incorrect processing, can prove impossible to section. These blocks are normally detected when being trimmed with a razor blade.
3. Incorrect clearance angle. Knives will not section if the clearance angle is too small; the block rubs against the back of the knife. If the angle is too large, the knife will tend to dig into the block.
4. Blunt knife. A blunt knife will often cut very bad sections rather than fail to section at all. This is usually due to prolonged final trimming. Replace the knife.
5. Microfeed rate too low. This is liable to happen with an ultramicrotome having a mechanical advance when the room temperature drops rapidly.
6. Unsuitable cutting speed. Sectioning at a cutting speed too high or low for the consistency of the block can result in a failure to section.
7. Microfeed at maximum extension. Ultramicrotomes with a thermal microfeed give no warning when the specimen arm is at its maximum extension. Ensure that the thermal microfeed is not accidentally left operating when it is not required. With instruments of the mechanical advance type it is good practice to check the amount of remaining travel of the specimen arm micrometer.

Should these adjustments be necessary, avoid wetting the face of the block when rematching it to the knife by removing a small quantity of water from the trough.

Skipping

The skipping of sections is another possible difficulty in ultrathin sectioning. Alternate sections are skipped by the ultramicrotome, and the sections that are cut are thicker than would be expected from the setting of the specimen micro-advance.

1. Soft blocks. Skipping sections is only one of the many problems caused by blocks which are too soft. Possible remedies are re-polymerization of the blocks or the use of a knife with a 35° edge angle.
2. Blunt knife. A blunt knife can cause the same problems.
3. Microfeed insufficient. If the block is insufficiently advanced at each cycle, only alternate, or even only every third section, may be cut. As a remedy, increase section thickness to the full extent of the microfeed immediately after cutting a thick section. The following sections should be thick ones; these can gradually be reduced in thickness by reducing the amount of specimen micro-advance.
4. Blocks containing areas of different hardness. Reorientate the block in its holder, or alternatively make the block face very small.
5. Knife clearance angle too great. A clearance angle that is too large can cause the knife to dig into the block face on alternate sections.

Scoring

Scores present another problem in sectioning; they are characterized by lines in the sections and on the block face. These are at right angles to the knife edge. Two types of score exist.

1. Large scores. These scores, sometimes large enough to cause the section to divide, can be caused by an inclusion in the block that breaks the knife edge, for example a chip of glass. These can sometimes be removed with a pair of forceps.

   Damage to the knife edge by careless handling or a badly made knife can also cause large scores.

2. Micro-scores. These fine scores which cover the whole block face are often only revealed by examination in the electron microscope. They are usually a result of cutting sections which are too thick when the face is being finally matched to the knife edge.

   In all cases of scoring the only solution is to use a new knife. Remove any foreign bodies lodged in the block face that could damage a new cutting edge.

Wetting

Wetting of the block face is another fault, and it is first noticed when cutting ultrathin sections. As the block disappears below the knife edge, water may be drawn to the very fine space between the back of the knife and the block face. This water is still present adhering to the block face at the start of the next sectioning cycle. With the block face in this condition it is impossible to cut ultrathin sections. This condition is invariably caused by the water in the floating-out trough being too high. The water wets the block before the edge of the knife makes contact with the lower edge of the block. Unless prompt action is taken to stop sectioning and remove the water from the face of the block with a filter paper, it will soak into the surface of the block and make sectioning very difficult until the wet resin is removed by trimming. This may mean the loss of important tissue.

When wetting occurs, stop sectioning. Remove the water from the block face with filter paper, and remove the bulk of the water in the floating-out trough. Very often a drop of water is noticed on the knife back; this is allowed to evaporate naturally before further sectioning is attempted.

Retract the block 2 μm to allow for any stray thermal advance during the pause in sectioning. Continue sectioning after rematching the block to the edge. Remember to have the water only just touching the cutting edge of the knife.

Chatter

Chatter is seen as a series of alternate light and dark bands in the section as it floats away from the knife edge. The presence of micro-chatter, which is only seen in the electron microscope, is indicated when the outline of tissue detail is seen on sections as they lie on the surface of the water. Chatter can be caused by any one of the following:

1. Soft block, blunt knife, loose knife and unsuitable clearance angle. These are the four faults common to most sectioning problems; they should be investigated and remedied if present.
2. Vibration. Vibration transmitted to the ultramicrotome (either by the operator touching the instrument, or from the bench or flooring) can cause chatter.
3. Badly shaped pyramids. A pyramid with ragged edges can cause chatter. The minute differences in width of block face in contact with the knife edge due to its irregular sides causes momentary changes in the cutting speed of the block with the fluctuation in cutting forces. Block faces in which the tissue only presents on one side of the face will almost certainly have chatter marks originating from the tissue/resin junction. A pyramid with regular smooth sides, none of which consists entirely of resin, is the answer.
4. Blocks with areas of different consistency. These blocks sometimes produce chattered sections as a result of the changing cutting speeds as the knife passes through areas of changing hardness in the block. There are two good solutions to this problem.

   The first is to persevere with the existing block face, turning it through 90° and sectioning in each plane until a full circle has been made in an attempt to find the best position. Occasionally a position is found using this technique that will give perfect results. As a last resort sections giving minimal chatter may have to be taken.

   The second solution for lessening chatter can be effected if one of the harder or softer areas is removed by reshaping the pyramid. This does not necessarily mean that one area will be lost because trimming the block to a step shape (Fig. 32) and sectioning the soft area first, followed by a retrimming to make the hard area prominent, can be successful.

Operator fatigue

Operator fatigue is another factor to be considered. Ultrathin

Fig. 32. A block fashioned to make the softer area prominent. Ultrathin sections are taken from this, and the block is retrimmed to make the hard area prominent for sectioning.

sectioning can rapidly tire the eyes and patience of the operator so that the smallest difficulty in sectioning appears to be insurmountable. In these circumstances changing to a less demanding task for a short period can often give the fresh approach necessary when cutting difficult blocks.

Holes

The final difficulty in ultrathin sectioning which will be discussed is holes in the sections. These may only be apparent when the section is examined in the electron microscope. A very occasional small hole can be of considerable use to the electron microscopist as it provides a built-in means of checking the astigmatism of the objective aperture. Certain tissues, for example adipose tissue containing numerous small fat droplets, tend to produce sections with holes which are difficult to eliminate.

5 Staining Ultrathin Sections

CARE OF GRIDS BEFORE AND AFTER EXAMINATION

To avoid contamination of the sections with dust, it is important to keep the grids in clean, closed containers at all times. Prior to staining, and before examination in the electron microscope, grids may be stored, sections uppermost, resting on a piece of clean filter paper in a Petri dish. As an aid to identification, the grid numbers should be clearly written on the filter paper.

After examination, each grid is placed in an empty gelatin capsule, one grid to each capsule. The capsules are labelled with the block identification number and stored with the original block for future reference or teaching purposes. Grids, carefully stored in this manner, show no deterioration on re-examination after several years.

A small plastic grid box marketed by LKB Ltd. has spaces for 100 grids standing on edge in indexed diamond-shaped holes. The container has a sliding lid to keep the grids dust free, and care must be taken when closing the lid to avoid cutting the grids in half if they are standing proud from their holes. A careful record of the position of each grid must be kept.

STAINING

Most tissues which have been fixed in osmic acid and embedded in methacrylate require no additional staining as the methacrylate sublimes in the electron beam, reducing the thickness of the section locally and increasing contrast. Although epoxy resins do sublimate to a certain extent, the contrast of osmic acid fixed tissues embedded in them tends to be very poor. Additional contrast has to be introduced for routine purposes. This increase in contrast is achieved by the use of heavy-metal salt solutions, loosely termed 'stains'. Unlike conventional stains, these solutions do not colour the tissue but deposit heavy-metal molecules onto the tissue to scatter or stop electrons so that they do not reach the image screen or photographic plate.

Stains can be applied to tissue blocks during fixation or dehyration. The more usual practice is to stain ultrathin sections on the grid.

Fixatives Which Also Stain

Concurrent with their fixing action, certain fixatives stain some tissue components. For example, osmic acid stains unsaturated lipids and phospholipids; potassium permanganate stains membranous structures of the cell and glycogen.

Stains Added to Dehydration Fluids

Phosphotungstic acid can be added to the dehydrating alcohols in order selectively to increase the contrast of connective tissue components. Normally 1-2 per cent of phosphotungstic acid is added to the higher alcohols during dehydration. Propylene oxide extracts phosphotungstic acid from blocks, and acetone should be substituted for it in the embedding procedure.

Uranyl acetate can be used in a similar manner to phosphotungstic acid. These methods may tend to make the tissue hard and difficult to section.

STAINING SECTIONS ON GRIDS

Alkaline Lead Solutions

Lead greatly increases the contrast of tissues embedded in epoxy resins. Probably the most commonly used stains for ultrathin sections are lead hydroxide, lead tartrate, and lead citrate. Lead hydroxide stains react with atmospheric carbon dioxide to form a precipitate of lead carbonate, which can cause heavy contamination of sections whilst they are staining. Because of the contamination problem, these stains have largely been superseded by Millonig's lead tartrate and Reynolds' lead citrate, both of which have excellent staining properties. At the same time, they produce little or no contamination of sections stained in them. Only the preparative details of Reynolds' lead citrate will be given as it is probably the more popular stain.

The poisonous nature of lead salts should not be forgotten when handling their solutions, and the appropriate precautions should be taken.

Reynolds' lead citrate

Prepare:

Solution A

Lead nitrate	1.33 g
Glass-distilled water	15 ml

Solution B

Sodium citrate	1.76 g
Glass-distilled water	15 ml

Allow the constituents of solutions A and B to dissolve completely in the water.

1. Add solution A to solution B dropwise, shaking all the time until completely mixed. A heavy white precipitate will form. Shake the solution vigorously at 5 min intervals for 30 min.
2. At the end of this mixing period, add 8 ml of 1 M-sodium hydroxide to dissolve the precipitate. The sodium hydroxide should be freshly prepared and free from sodium carbonate to avoid the formation of a precipitate in the final stain.
3. Dilute the stain to a total volume of 50 ml with distilled water. The stain should be clear. Store in a clean bottle. A deposit slowly forms on the bottom of the bottle as the stain ages, but this does not interfere with its staining properties.

Staining times with lead salts

Millonig's lead tartrate and Reynolds' lead citrate rapidly stain tissue in epoxy resins. The time for staining with each batch of stain seems to vary, but it normally takes 1-3 min. Overstaining is indicated by a scattering of a dense, granular precipitate covering the section. The size of individual grains of the contamination depends on the degree of overstaining. Large pieces may be present at low magnification in the case of gross contamination, whilst a very fine contamination may only be visible at higher magnifications. This contamination can be removed by immersing the grid in 1 per cent acetic acid in distilled water for a few seconds, followed by a wash in distilled water. This treatment removes all traces of the lead stain. Therefore, the sections must be restained using an appropriately shorter time.

Tissues may vary considerably in their affinity for stains, and it has been found that there is a certain degree of variation from block to block with the same stain. The exact amount of staining required, therefore, is a matter for trial and experience. If a section appears to lack contrast when first examined in the electron microscope, it may be usable at higher magnifications. Restaining of a section that has been examined in the electron microscope is sometimes found not to be very effective. The reason for this is not clear, but it is frequently preferable to cut another ultrathin section and to stain this for a longer period.

When examining a small area of a section at very high magnifications, a local build-up of contamination on the section tends to reduce the contrast.

Procedure for Staining Ultrathin Sections

1. Take a clean piece of paraffin wax, approximately 4 x 4 x 1 cm, and scoop a shallow hollow 1 cm across on one of the larger sides. This is to be the upper surface of the staining block.
2. Place the piece of wax in one half of a clean Petri dish, and surround the wax with water. This reduces the rate of evaporation of the stain and thus slightly reduces the amount of contamination.

3. Place two or three drops of the lead citrate stain in the wax hollow. Remove the stain from its stock bottle with a new disposable or clean Pasteur pipette, being careful to expel air from the teat before dipping the tip of the pipette under the surface of the stain. This prevents the disturbance of sedimented crystals or precipitate. If the stain bottle is kept on a shelf just above eye level, the very thin film of crystals on the surface of the stain can be easily seen and avoided with the tip of the pipette. All stains should be crystal clear before use. Even a slight haze in the solution itself will normally cause very heavy contamination of the section.
4. The unstained dry grid is placed onto the surface of the stain, sections facing downwards.
5. Cover the Petri dish with its lid (Fig. 33).

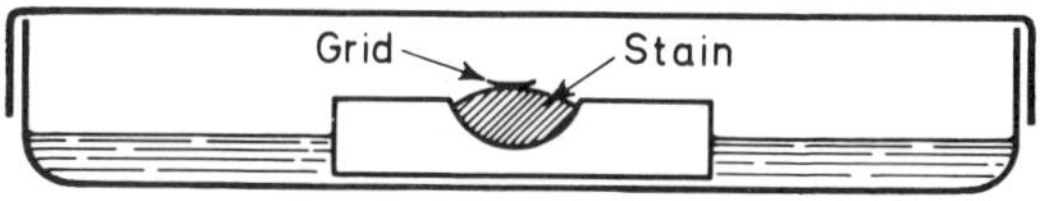

Fig. 33. A Petri dish being used as a staining dish for grids, it contains a piece of wax surrounded by water. The stain is placed in a hollow in the wax, with the grid floating on the stain.

6. After sufficient staining time has elapsed, remove the grid by using a pair of clean, fine pointed forceps to catch hold of its rim.
7. Wash the grid for a few seconds in a stream of glass-distilled water; hold the grid over a clean glass beaker in order to catch it should it slip from the forceps. If a grid is dropped on the floor, it should be rejected as it is practically impossible to remove the contamination.
8. Remove the surplus water from between the points of the forceps with a piece of clean filter paper and from the surface of the grid by touching its edge.
9. Allow the grid to dry naturally in a dust-free environment whilst being held in the points of the forceps (Fig. 34). The grid is now ready for examination in the electron microscope.

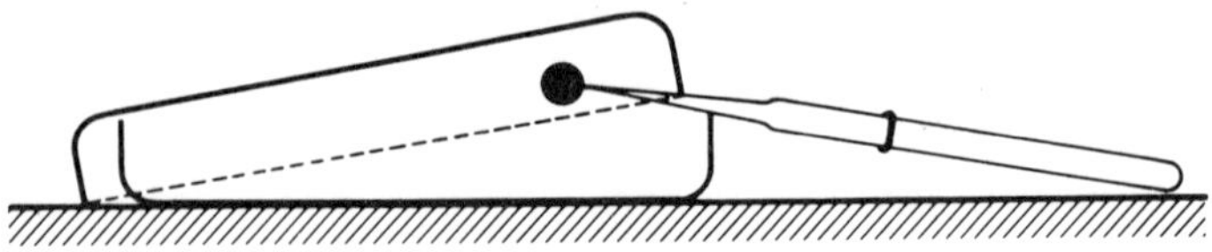

Fig. 34. A grid held in forceps and covered by a Petri dish whilst it is drying.

Double Staining

The intensification of contrast by double staining with uranyl acetate and lead citrate is carried out as follows:

1. Using the technique described above, stain the sections with a saturated aqueous solution of uranyl acetate. Biological tissues embedded in Epikote normally require 3-5 min.
2. When the grid has been washed and dried, repeat the technique using Reynolds' lead citrate stain for 1-3 min. The grid is again washed and dried as described above and is now ready for examination.

Hints and Tips

Carelessness at this stage means needless repetition of painstaking work. The contamination needed to ruin a section is invisible to the naked eye and is normally only apparent on examination in the electron microscope. Confidence in the whole technique is only achieved with practice. The following conditions are mandatory:

1. A clean part of the bench must be selected as the staining area. This area should preferably be away from the sites of dust-producing activities such as sawing and filing of blocks.
2. Scrupulously clean glassware should be used to make up stains. Clean bottles should be used for storing the stains.
3. Use clean forceps to pick up grids.
4. The stains should be made from good quality chemicals and glass-distilled water.
5. Centrifuge any suspect staining solutions as filtration may introduce more contamination than it removes.

Uranyl Acetate

A saturated aqueous solution of uranyl acetate may be used to stain sections to give a useful increase in the contrast of DNA and tissue proteins. Although this stain gives a useful increase in the general contrast of sections, it fails to stain cell membranes adequately. Uranyl acetate stain may be used with advantage prior to staining with an alkaline lead stain when sections with a high degree of contrast can be obtained. This technique is known as double staining.

Preparation of the stain

1. Dissolve small amounts of uranyl acetate in warm glass-distilled water until no more will dissolve. A saturated solution is approximately 8 g per cent at room temperature.
2. This solution is cooled at 4° C overnight, and the crystalline deposit which forms is removed by centrifugation.
3. When kept in a clean bottle, this stain is usable for several months. During this period, crystalline deposits may occur on the bottom of the bottle and on the surface of the stain. These deposits do not detract from the solution's staining properties, and it is only necessary to remove them by centrifugation when it is impossible to remove the clear liquid from the bottle without contamination.

The staining time is found by trial and error. Three to five minutes normally works very well with tissue embedded in Epikote resin, but occasionally periods of up to 30 min may be necessary.

Phosphotungstic Acid

Various concentrations, for example 5 per cent or 10 per cent phosphotungstic acid in glass-distilled water, can be used to stain sections. The exact time necessary must be found by trial and error. Collagen, elastic fibres, and mucopolysaccharides are stained by phosphotungstic acid, but cell membranes are not.

6 Problem Tissues

If the methods described in previous chapters are followed, it should be possible to prepare satisfactory ultrathin sections from most soft tissues as a routine procedure. Some tissues, for example skin, blood cells and bone, require special techniques, and experimentation is necessary in order to obtain good results.

SKIN

Ultrathin sections of mature human skin showing the correct relationship between the various cell layers without breaks appearing between them can be very difficult to obtain. These breaks are probably caused by the different sectioning requirements of the dermal and epidermal layers with their differences in hardness and texture within one block. To obtain correct orientation, small pieces of skin should be embedded in moulds that allow manipulation of the block. Moulds made from aluminium foil which allow the correct plane of section through the cell layers to be obtained without further embedding are very useful (Fig. 35).

If it is necessary to preserve the dermal/epidermal junction intact, the area selected for ultrathin sectioning should be made as small as possible, consistent with obtaining the desired information. Larger sections can be made of the dermis or epidermis separately. The best results are usually obtained by positioning the block in the ultramicrotome to cut the dermal layer first, followed by the epidermis, using a slow cutting speed.

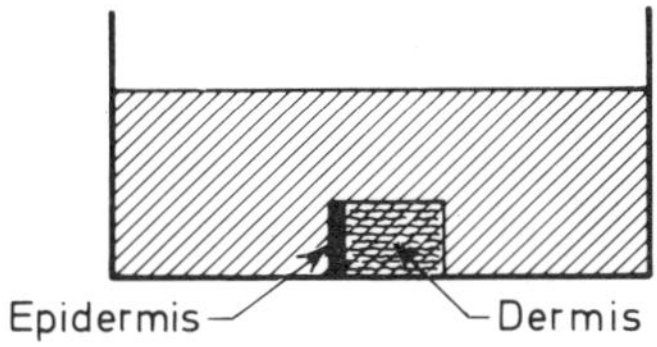

Fig. 35. Aluminium foil mould, used for embedding blocks of skin to obtain a section through both the dermis and epidermis.

BLOOD CELLS

Before any procedure is undertaken the problem should be discussed in detail with a haematologist in order that the blood may be taken into an appropriate anticoagulant and the various cellular elements, for example platelets, leucocytes or erythrocytes, separated. The method of fixation should also be discussed as it is extremely easy to introduce artefacts inadvertently into this type of specimen.

Blocks from which good ultrathin sections can be obtained and which contain well fixed platelets can be made by the method described below. It is possible to substitute other blood cells for platelets in this technique if sections of these are required.

Embedding Methods for Platelets

1. Prepare platelet-rich plasma from whole blood.
2. Mix 1 ml of this plasma with 1 ml of Palades 2 per cent osmic acid fixative in a small centrifuge tube. Allow the platelets to fix for 5 min at 37° C. The plasma, fixative and centrifuge tubes are prewarmed to 37° C before use. Abnormally shaped platelets occur if fixation is carried out at 4° C; this is the temperature at which osmic acid is normally used (O'Brien and Woodhouse, 1968).
3. Centrifuge the platelets for 2 min at 3000 rev./min to form a pellet in the bottom of the tube. To avoid repeated centrifugation and possible tissue damage, the platelet pellet is left in the bottom of the tube for the whole of the embedding process. The pellet is only removed for the final stage of embedding in the gelatin capsule.

4. Without disturbing the pellet, pour off the osmic acid, and replace it with distilled water for 1 min.
5. Give two changes of 3 min each in 70 per cent alcohol.
6. Give two changes of 3 min each in 90 per cent alcohol.
7. Give two changes of 10 min each in absolute alcohol.
8. Give two changes of 10 min each in propylene oxide.
9. Give one change of equal parts of propylene oxide and Epikote resin mixture for 30 min. To avoid disturbing the platelets on the botom of the tube this mixture should be made separately in a small polyethylene bottle.
10. Change to undiluted resin for 30 min.
11. Remove the platelets from the tube with a small round-ended spatula, trying to keep the platelets as a pellet. If a larger than normal number of platelets were in the original sample, the pellet can be cut into two or three pieces with a razor blade in order to make blocks with overall dimensions of approximately 1 mm.
12. Drop the pellet or piece of pellet onto the resin surface of an embedding capsule which has been filled with fresh resin.
13. Place the block in an incubator at 60° C to polymerize the resin. The platelets will sink to the bottom of the capsule.

BONE

Hard mineral salts, which form an integral part of bone, make the preparation of ultrathin sections from this tissue very difficult. Glassknife cutting edges are quickly broken down; use a diamond knife, if possible, to section this tissue.

COLLAGEN

Collagenous tissue is often extremely dense, and it may be necessary to extend the later stages of the embedding process to allow full penetration. On examination in the electron microscope, it is sometimes found that small holes are present at the sites of individual collagen fibres. Areas in a block can usually be found which section satisfactorily.

LIPIDS

There are many different lipids some of which present no great difficulty to the ultramicrotomist; others, especially large deposits of neutral fat, can cause sectioning problems. This is usually manifested as holes which replace the lipids. The problem is extremely difficult to overcome, and the electron microscopist often has no alternative but to examine the tissue remaining between the holes. Sometimes the use of a different fixative or embedding medium may help.

7 Histochemistry

Histochemical reactions for certain enzymes have been adapted for use in electron microscopy. The methods most successfully used have been those relying on the deposition of heavy-metal salts as a reaction product at the site of the enzyme; for example, lead phosphate is deposited at sites of acid phosphatase activity in the Gomori type reaction.

DEMONSTRATION OF ACID PHOSPHATASE (MODIFIED GOMORI)

In this method, the enzyme acid phosphatase produces phosphate from sodium β-glycerophosphate at pH 5.0. The phosphate formed combines with lead ions present in the substrate to form insoluble lead phosphate at the site of the enzyme activity. For light microscopy this is converted to black lead sulphide by the action of ammonium sulphide. Electron microscopy does not require this final conversion to lead sulphide as lead phosphate is electron dense and visible in the electron microscope.

Prepare

1. 0.05 M-Acetate buffer

Solution A. 0.1 M-Acetic acid

Glacial acetic acid (AnalaR)	2.85 ml
Glass-distilled water	500 ml

Solution B. 0.1 M-Sodium acetate solution

Sodium acetate hydr. (AnalaR)	6.8 g
Glass-distilled water	500 ml

To 59 ml of solution A, add 141 ml of solution B and 200 ml of glass-distilled water. The pH of the buffer is adjusted to 5.0 by the addition of small amounts of either solution A or solution B. The pH adjustments are monitored using a glass-electrode pH-meter.

2. Dissolve 0.5296 g of lead nitrate (AnalaR) in 400 ml of 0.05 M acetate buffer. This stock solution keeps well at 4° C for several months.
3. Dissolve 3 g of sodium β-glycerophosphate in 100 ml of glass-distilled water. This reagent should be of the type specified to contain a low percentage of the alpha form (Taab laboratories). The pH of this solution should now be checked and adjusted to 5.0 by the addition of drops of glacial acetic acid, monitored with a pH meter. Most batches of glycerophosphate are approximately pH 7.0-9.0 when first made into solution. At first several failures occurred with this technique due to the formation of a heavy precipitate on the sections during incubation at 37°C. if the pH of the glycerophosphate solution is adjusted as described, very little if any precipitate will form in the substrate after several hours' incubation at 37° C. The substrate can often be used when freshly made up without the preincubation and filtering described in most methods. An occasional batch of substrate may produce a very fine precipitate after several hours' incubation. This can be removed by filtration before using the solution.
4. Substrate. Add 10 ml of 3 per cent sodium β-glycerophosphate solution at pH 5.0 to 100 ml of stock lead nitrate solution at pH 5.0. Place in an incubator at 37° C, and allow the solution to reach this temperature before use. If a precipitate forms, the solution should be filtered.

Method

1. Cut 10 μm-thick frozen sections from tissue fixed in glutaraldehyde or formaldehyde fixatives. The tissue should be washed in the sucrose/cacodylate buffer described in Chapter 1 for at least twice as long as the fixation time before cutting sections. In practice this time need not exceed 24 h.

2. Place the sections as they are cut in sucrose/cacodylate buffer at 4° C until required for use.
3. To find the most suitable incubation time for electron microscopy, transfer ten of the sections directly from the buffer to the substrate, pre-warmed to 37° C.
4. Incubate the sections in the substrate at 37° C. Remove one section from the substrate with a clean glass rod after 5 min.
5. Wash in several changes of glass-distilled water.
6. Immerse the section in 10 per cent ammonium sulphide solution in distilled water for 30 s.
7. Wash in distilled water.
8. Flatten the section onto a clean glass microscope slide. Mount in a water mounting medium, and cover with a coverglass. Examine the section under a microscope for a black reaction product.

Slides are successively made at 5 min intervals, using the same technique to follow the build-up of the reaction product in the section. The shortest incubation time, giving a well localized black deposit in lysosomes with no sign of nuclear staining, is most suitable for electron microscopy. Acid phosphatase present in lysosomes of the proximal tubules of rat kidney can usually be demonstrated after 12-20 min incubation.

Sections for electron microscopy

1. Incubate frozen sections cut at 30 μm in the substrate for the determined time.
2. Remove the sections from the substrate on completion of the incubation. Give three washes each of 30 s in glass-distilled water.
3. Post-fix the sections by placing them in Palade's 2 per cent osmic acid fixative for 2 h at 4° C.
4. Give the sections a quick rinse in distilled water.
5. Dehydrate, giving one minute in each of the alcohols.
6. Give two changes each of 5 min in propylene oxide.
7. Place in equal parts of propylene oxide and Epikote resin mixture for 15 min. Do not make this mixture in the bottle with the sections as they are very brittle at this stage.

8. Immerse the sections in undiluted resin for 20-30 min.
9. Fill a 10 mm-deep mould with resin mixture to a depth of 5 mm. Place the resin-infiltrated sections onto the surface of the resin in the mould, allowing 5-10 mm between each section. The sections will sink to the bottom of the mould after 30 min incubation at 37° C. They can be flattened and separated from one another with a fine glass rod. Sections which will not flatten completely in this way may be flattened by having a small block of solid resin with one flat side, placed over them as a weight. This block of resin will become completely homogeneous with the resin surrounding the section and not cause any difficulty when sectioned. The mould is incubated at 60° C to polymerize the resin.

From one of these embedded sections cut a 1-2 μm-thick section, mount it on a glass microscope slide, and allow it to flatten and dry in the usual manner. When completely dry, cover the section with a mixture of equal parts of ammonium sulphide solution and distilled water. After one minute, wash the slide in distilled water, and allow it to dry. Mount in a synthetic mounting medium, cover with a coverglass, and examine in the light microscope under phase contrast. Areas of lead phosphate deposition can be identified as black, highly refractive masses. These black areas should remain visible when examined by ordinary light microscopy. Trim the blocks to a size suitable for ultrathin sectioning. Make two ribbons of ultrathin sections, and mount on separate grids. Stain one with lead citrate, and leave the other unstained. Examine each grid in the electron microscope. Positive areas appear as localized black, granular precipitates occurring in the same position in the stained and unstained sections (Fig. 36).

Controls

In order to eliminate false positive results, the following controls should be observed.

1. Incubate sections in the complete substrate with the addition of 0.01 M-sodium fluoride (0.021 g/50 ml). This inactivates

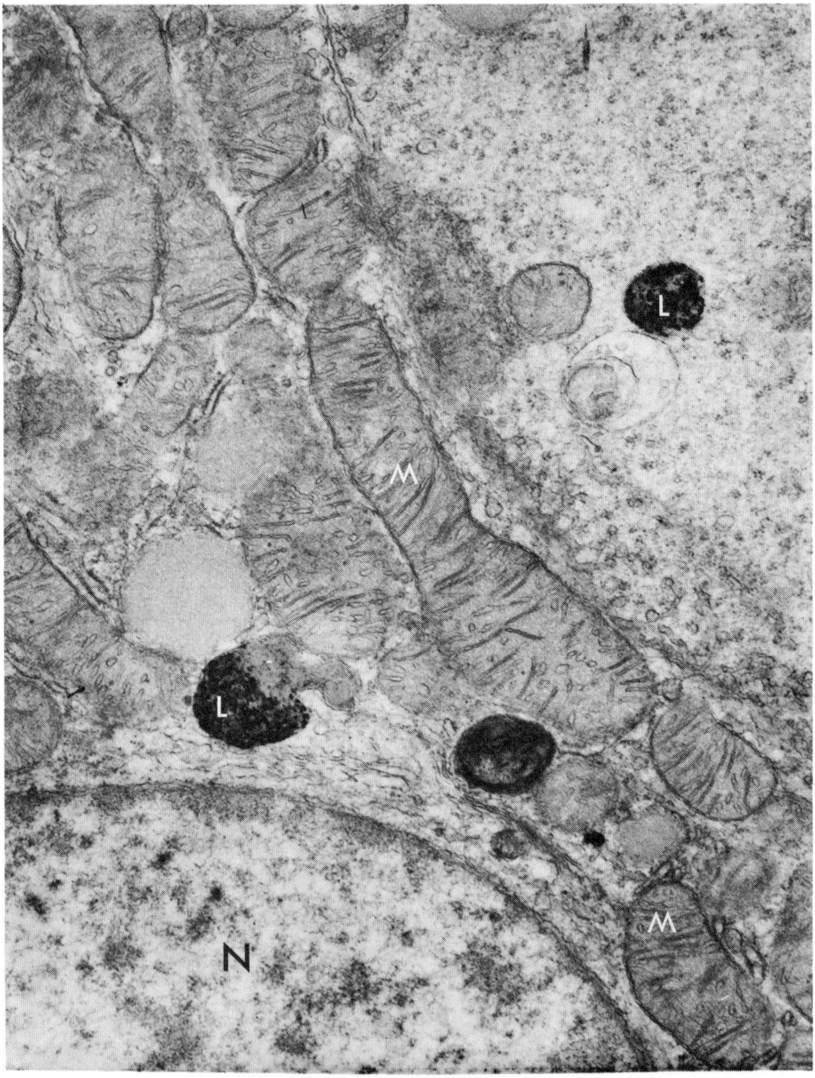

Fig. 36. The demonstration of the enzyme acid phosphatase in lysosomes, L, by the Gomori reaction. Notice that the site of the enzyme, shown as a black deposit of electron-dense lead phosphate, is well localized, with no reaction product or contamination visible in either the cell cytoplasm or nucleus, N. M indicates mitochondria.

Tissue: Rat kidney fixed in glutaraldehyde, embedded in Epikote resin and the section stained with lead citrate. Magnification: × 28 000.

acid phosphatase; no reaction product should be visible in the light or electron microscopes.

2. Incubate sections in the substrate without the addition of sodium β-glycerophosphate. Omission of this compound should give a negative result.

Very occasionally, false positive results have been obtained with the Gomori-type acid phosphatase technique in which non-reacting electron-dense intracytoplasmic organelles were stained very heavily with the lead nitrate in the absence of glycerophosphate. The differentiation of acid phosphatase containing lysosomes from these dense bodies was extremely difficult.

ALKALINE PHOSPHATASE

Calcium phosphate is formed by the enzyme alkaline phosphatase from sodium β-glycerophosphate in the presence of calcium chloride at pH 9.0. The calcium is then replaced by cobalt which is rendered visible for light microscopy by the action of ammonium sulphide to form black cobalt sulphide. The final stage is unnecessary for electron microscopy as cobalt is electron dense.

Prepare

Solution A

Sodium barbitone	6.1 g
Calcium chloride (anhyd.)	1.2 g
Magnesium sulphate ($MgSO_4.7H_2O$)	0.5 g
Distilled water	1000 ml

This solution is still usable after storage at 4°C for several months.

Solution B

Sodium β-glycerophosphate	1.0 g
Distilled water	100 ml

This solution is freshly prepared before use.

Substrate

Solution A	50 ml
Solution B	30 ml

Adjust the pH of the substrate to 9.5 by the addition of 0.1 M-hydrochloric acid or 0.1 M-sodium barbitone. Use a glass electrode pH-meter to monitor these adjustments.

Method

1. Cut 30 μm frozen sections from glutaraldehyde- or formaldehyde-fixed tissue. The tissue should be washed in the sucrose/cacodylate buffer described in Chapter 1 for at least twice as long as the fixation time before cutting the sections. In practice this need not exceed 24 h.
2. Place the sections as they are cut into sucrose/cacodylate buffer at 4° C until required for use.
3. Filter the substrate, and warm to 37° C before use.
4. Incubate several sections in the substrate for 20 min.
5. Remove the sections from the substrate with a clean glass rod.
6. Wash in five changes of glass-distilled water, giving one minute in each.
7. Place the sections in a 2 per cent solution of cobalt nitrate in glass-distilled water for 5 min.
8. Wash in five changes of glass-distilled water, giving 2 min in each.
9. Post-fix the sections in 2 per cent osmic acid for 2 h at 4° C. Embed the sections in Epikote resin by the technique described for acid phosphatase.

   Cut a 1-2 μm section from an embedded 30 μm section, mount it on a glass microscope slide, and allow it to flatten and dry. When completely dry, cover the section with a mixture of equal parts of ammonium sulphide solution and distilled water. After 1 min wash the slide and allow to dry. Mount in synthetic mounting medium.

   Sites of alkaline phosphatase activity are indicated by black deposits and frequently occur along membranes. Select an area of the block showing a positive result for ultrathin sectioning, and make two grids each with a ribbon of sections from the same block face. Stain one with lead citrate and leave the other unstained.

A black, granular precipitate is seen at sites of alkaline phosphatase activity. This should be visible in the same position in both the stained and unstained grids.

Controls

1. Incubate sections without the sodium β-glycerophosphate. These sections should give a negative result.

   Tissues containing calcium deposits give a false positive reaction with this method. The calcium can be removed by treating sections with 0.1 M-citrate buffer at pH 5.0 for 5-10 min before incubation in the substrate. This process is not detrimental to the reaction of the enzyme.
2. A useful tissue for positive control is the brush border of the proximal convoluted tubule of the kidney (Fig. 37).

HISTOCHEMICAL METHODS ON RESIN-EMBEDDED TISSUE FOR EXAMINATION IN THE LIGHT MICROSCOPE

Some of the methods used in conventional histology for the demonstration of metallic complexes in paraffin wax sections can be applied to sections from tissues embedded in epoxy resin. The following methods have been used to demonstrate haemosiderin and calcium salts in 2 μm Epikote sections and are useful when selecting an area for ultrathin sectioning from specimens containing small deposits of these.

Haemosiderin

Perl's Prussian blue method can be used to demonstrate haemosiderin.

1. A flat 2 μm section of tissue embedded in Epikote is required. The section should adhere well to the slide on which it is mounted.

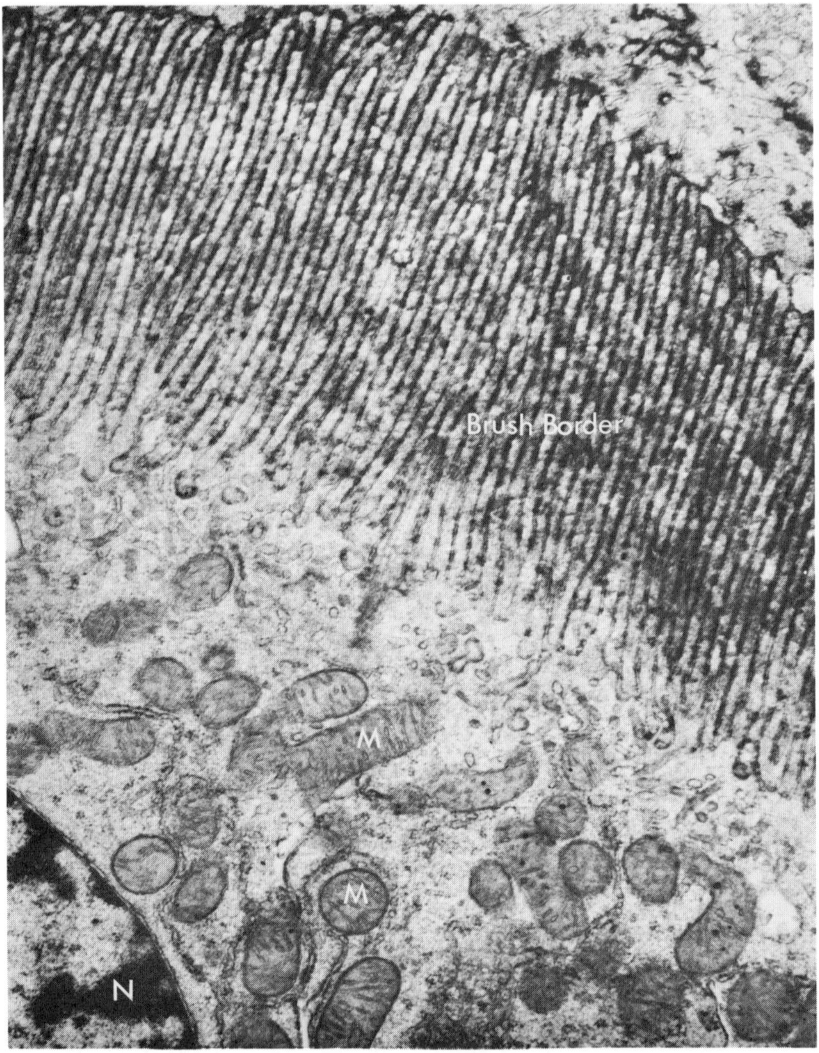

Fig. 37. The site of the enzyme alkaline phosphatase can be seen as a dense black deposit along the membranes of the brush border. The method used substitutes electron-dense cobalt for calcium in calcium phosphate formed by the enzyme from sodium β-glycerophosphate in the presence of calcium chloride at pH 9.0

Tissue: Rat kidney proximal tubule fixed in buffered glutaraldehyde, embedded in Epikote resin and the section stained with lead citrate. Magnification: × 28 000.

2. Immerse the slide in the following solution for 15-20 min at 45° C in a Coplin jar:

   10 per cent potassium ferrocyanide in distilled water,
   10 per cent hydrochloric acid (AnalaR) in distilled water.

   For use, mix equal parts of these solutions.
3. At the end of the reaction time, wash the section in distilled water. Do not wash it too vigorously as it may become detached from the slide.
4. After drying the slide, mount the section in a synthetic medium.
5. A positive control should be run in parallel.

Result

Haemosiderin: deep blue

The sections are not normally counterstained as this masks very small deposits. Examination in the phase contrast microscope may facilitate the orientation of tissues, but the presence of the blue colour of haemosiderin must be confirmed by ordinary light microscopy.

Calcium

The following methods have been found to work on calcium salts in tissues embedded in Epikote resin. Sections cut at a thickness of 2 μm are used to select areas containing small calcium deposits suitable for ultrathin sectioning.

Alizarin red S (McGee-Russell, 1958)

Add 5 per cent ammonia solution dropwise until the pH of a 2 per cent solution of alizarin red S in distilled water is 4.1-4.3. The pH of this solution is critical, and a pH-meter should be used to monitor the addition of the ammonia.

Method

1. Cover a 2 μm-thick resin section attached to a glass microscope

slide with one or two drops of alizarin red S stain. Place a coverglass over the section to produce a thin layer of stain between the slide and coverglass.
2. Observe the section under a microscope using a low-power objective. Deposits of calcium gradually stain a bright orange to red colour, developing maximum coloration in a few minutes. Prolonged staining will produce a diffuse reaction.
3. When a strong staining reaction is present, wash off the stain and coverglass with acetone. Blot the section with a clean filter paper, and place the slide in a 60° C incubator to dry.
4. Mount the section in a synthetic mounting medium.

Result
Calcium deposits: bright orange/red colour

Von Kossa (modified)

1. Place a microscope slide with attached resin section in a 2 per cent solution of silver nitrate in glass-distilled water. Stain for 30 min in the dark.
2. Wash in glass-distilled water. Give several changes of one minute each.
3. Develop in a freshly prepared 0.5 per cent solution of photographic hydroquinone in distilled water for 5 min.
4. Rinse in distilled water, dry the section at 60° C, and mount.

Result
Calcium deposits: black

This method is not specific for calcium. Phosphates and carbonates demonstrated by this technique are presumed to be combined with calcium at the blackened sites.

8 Support Films

In order to examine the thinnest ultrathin sections, viruses, and particulate specimens in the electron microscope, the copper grid must be covered with a very thin film of electron translucent material approximately 5-20 nm thick. This film acts as a support over the spaces in the grid and may be made from carbon or formvar, which is a plastic material.

Carbon films have greater mechanical strength and more stability in the electron beam than those made from formvar. They can usually be made thinner than formvar films and tend to be more useful for high resolution work. However, formvar films can be made more quickly than carbon films and tend to be less fragile.

VACUUM-EVAPORATED CARBON FILMS

Requirements

1. A vacuum coating unit capable of evaporating carbon from a carbon arc.
2. A supply of carbon rods of suitable diameter for the evaporation apparatus. These rods should be of spectrographically pure carbon.
3. Clean glass microscope slides.
4. A number of fine mesh grids.
5. A one per cent solution of Teepol or liquid surface-wetting agent in distilled water.

Method

1. Immerse a clean glass microscope slide in the Teepol solution holding it with a pair of forceps.

2. Remove the slide, and stand it on a clean filter paper in a vertical position. Allow the slide to drain and dry completely. It is now covered with a thin film of release agent.
3. Hold the slide by its edges, and place it on the work chamber floor of the vacuum coating unit. Set up the carbon arc in accordance with the manufacturer's instructions at a distance of approximately 100 mm from the surface of the slide. At the side of the slide, place a piece of white porcelain with one drop of diffusion pump oil on its surface. This acts as an indicator of the thickness of the carbon film.
4. Replace the work chamber cover, and evacuate the air.
5. When the working vacuum is reached, actuate the carbon arc for approximately half a second. Examine the porcelain in the work chamber. It will be seen that the area covered by the oil will still appear to be white, surrounded by a brown deposit. As an indication of carbon thickness, this colour should be very light brown. Extremely pale colours should be given very light additional coats of carbon by using short bursts of the arc. A heavy dark brown colour indicates a film which is too thick to be used and is best discarded.
6. When a satisfactory film has been made, remove the slide from the unit, holding it by its edges.
7. Use a fine pointed needle to divide the carbon film into small squares; each one should be sufficient to cover a grid.
8. Fill a large glass beaker to the brim with distilled water. Remove any dust from the water by pulling a clean filter paper across the surface.
9. Breathe on the carbon film. This will produce condensation along the scribed lines.
10. Dip the slide into the water at an angle of 45° to the surface; the squares of carbon film should float separately onto the surface of the water.
11. Dip a grid held in watchmaker's forceps under the surface of the water, and collect a square of film onto it by bringing the grid up under the film held at an angle of 45°.
12. When the grid is dry, place it on a clean filter paper in a Petri dish. The grids may be stored in these dishes until required for use. They will keep for quite a long time if left undisturbed,

but a sample should be checked from time to time to ensure that a supply is always available.

FORMVAR FILMS

These plastic films are made by the deposition of a thin layer of formvar onto a clean, flat surface from a solution by evaporation of the solvent.

Requirements

1. 0.15 per cent solution of formvar in 1,2-dichloroethane.
2. Clean glass microscope slides.
3. A large glass beaker.
4. Grids.

Method

1. Dip a clean glass microscope slide into the formvar solution for 30 s.
2. Remove the slide, and allow it to drain vertically on a piece of clean filter paper.
3. When it is dry, cut the film around its edges and across the middle with a razor blade. The film is not visible at this stage, and the cuts are made where the film is assumed to be.
4. Fill the beaker with distilled water, and float a filter paper across the surface to remove the dust.
5. Breathe onto the film. Dip the slide with its film of condensation into the beaker. Holding it an angle of 45° to the surface of the water, first immerse one corner so that the film will lift off the slide. When the slide is immersed up to the line, cut across it; the film will detach and float freely on the water. The slide is now removed.

   Sometimes the formvar film does not detach itself from the slide as the slide may have been cleaned too thoroughly. In practice, if slides are taken from a new box and dust particles brushed away, they will be found to be sufficiently clean.

6. The film is easily seen if the surface of the water is illuminated at an angle from one side.
7. Carefully place grids, matt side down, onto the formvar film. It is usually possible to place 15-20 grids on the film, but areas containing obvious imperfections such as holes or thick areas should be avoided.
8. When sufficient grids have been placed on the film, drop a piece of filter paper larger than the formvar film over it. Pick up the filter paper by one corner with a pair of forceps. The formvar film will adhere to the paper.
9. Place the filter paper, formvar film uppermost, in a Petri dish. Prop the lid of the dish up on the lip of the dish bottom to allow the paper to dry slowly.
10. The grids, complete with formvar film, can be detached from the filter paper by running a sharp needle around the edge of the grid to break the film holding it to the paper. Do not make formvar film on a humid day as the films produced will tend to contain numerous holes.

Both support films described are easily made and can be used with confidence for most techniques for which a support film is required.

References

GLAUERT, A. M. and GLAUERT, R. H. (1958) *J. biophys. biochem. Cytol.* **4**, 191

GONZALEZ, AQUILOR F. and DE ROBERTIS, E. (1963) *Neurology* **13**, 758

LUFT, J. H. (1956) *J. biophys. biochem. Cytol.* **2**, 799; (1961) **9**, 409

McGEE-RUSSELL, S. M. (1958) *J. Histochem. Cytochem.* **6**, 22

O'BRIEN, J. R. and WOODHOUSE, M. A. (1968) *Exp. biol. Med.* **3**, 90-102 (Karger, Basle and New York)

PALADE, G. E. (1952) *J. exp. Med.* **95**, 285

Suggestions for Further Reading

ATLASES OF ULTRASTRUCTURE

FAWCETT, D. W., *An Atlas of Fine Structure,* W. B. Saunders Co., Philadelphia and London

RHODIN, J. A. G., *An Atlas of Ultrastructure,* W. B. Saunders Co., Philadelphia and London

THREADGOLD, L. T., *The Ultrastructure of the Animal Cell,* Pergamon Press, London

TONER, P. G. and CARR, K. E., *Cell Structure,* E. and S. Livingstone, Edinburgh and London

TECHNICAL BOOKS

KAY, Desmond H., Editor, *Techniques for Electron Microscopy,* Blackwell Scientific Publications, Oxford

PEASE, Daniel, C., *Histological Techniques for Electron Microscopy,* Academic Press, New York and London

Suppliers of Equipment

MATERIALS FOR TISSUE PREPARATION AND ULTRAMICROTOMY

TAAB Laboratories, 52 Kidmore End Road, Emmer Green, Reading, Berkshire	Fixatives, embedding media, grids, and TAAB polyethylene capsules
Gurr, G. T., Searle Scientific Services, Coronation Road, Cressex Industrial Estate, High Wycombe, Buckinghamshire	Fixatives, embedding media, grids, and BEEM polyethylene capsules
LKB Ltd., 232 Addington Road, South Croydon, Surrey	Knifemaker, glass, and general ultramicrotomy suppliers
Solmedia, 31 Orford Road, London E.17	Glass embedding vials

ULTRAMICROTOMES

Cambridge Scientific Instruments, Histon Road, Cambridge	Cambridge Huxley Ultramicrotome

LKB Ltd. 232 Addington Road, South Croydon, Surrey	LKB Ultrotome
British American Optical Co., Instrument Group, 266 Bath Road, Slough, Buckinghamshire	Reichert Ultramicrotome
V. A. Howe & Co. Ltd. 88 Peterborough Road, London S.W.6	Porter-Blum Ultramicrotome

Index